German Starfighters

The Story in Colour: Introduction and Units

Volume 1

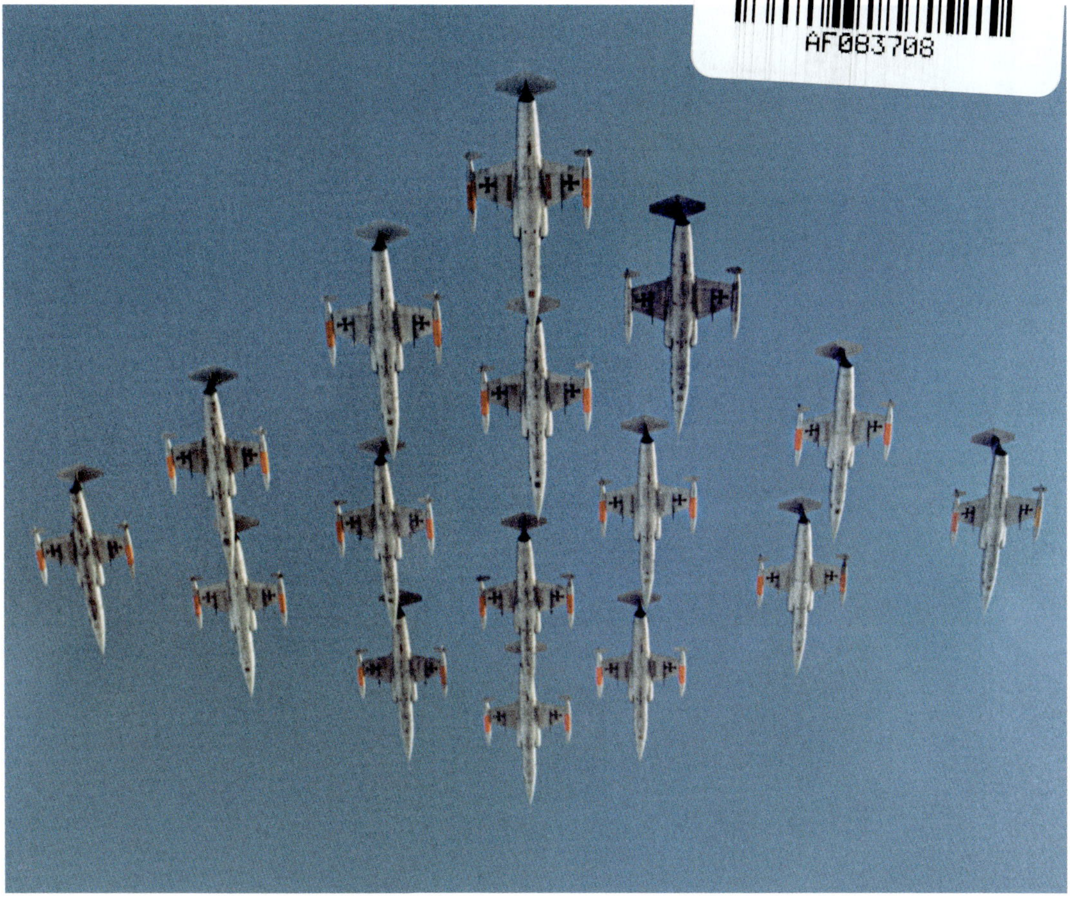

KLAUS KROPF

Key Books

HISTORIC MILITARY AIRCRAFT SERIES, VOLUME 24

Front cover image: Four F-104G Starfighters of Jagdbombergeschwader 32 in close formation. (Reinhart Blume)

Back cover image: F-104G Starfighter of Jagdbombergeschwader 34 passing the Alps on a sunny winter's day. (Karl Ascherl)

Contents page image: TF-104G Starfighter of Waffenschule 10 during a formation take-off at Jever Air Base (AB). (Ralf-Dietmar Pilawa)

Published by Key Books
An imprint of Key Publishing Ltd
PO Box 100
Stamford
Lincs PE9 1XQ

www.keypublishing.com

The right of Klaus Kropf to be identified as the author of this book has been asserted in accordance with the Copyright, Designs and Patents Act 1988 Sections 77 and 78.

Copyright © Klaus Kropf, 2022

ISBN 978 1 80282 474 2

All rights reserved. Reproduction in whole or in part in any form whatsoever or by any means is strictly prohibited without the prior permission of the Publisher.

Typeset by SJmagic DESIGN SERVICES, India.

Contents

Acknowledgements 4

Introduction 5

Foreword 6

Chapter 1   First Steps 8

Chapter 2   Starfighter Production 13

Chapter 3   Waffenschule 10 (WaSLw 10) – Weapon School 10/Training Wing 18

Chapter 4   Jagdbombergeschwader (JaboG) – Fighter-Bomber Wings 30

Chapter 5   Jagdgeschwader (JG) – Fighter Wings 59

Chapter 6   Aufklärungsgeschwader (AG) – Reconnaissance Wings 69

Chapter 7   Marinefliegergeschwader (MFG) – Naval Air Wings 76

Chapter 8   Kommando F-104 – F-104 Holding Unit 89

Abbreviations and Ranks 93

Bibliography 95

Acknowledgements

The production of this book would not have been possible without the help of many former comrades of the Luftwaffe and Naval Air Wings, as well as some civilian supporters. The search for colour images was not always easy. It was time consuming and required a lot of patience and intensive research, especially with regard to the 1960s. More than 400 colour pictures now provide a retrospective of the history of Starfighter operations in the German squadrons from 1961 to 1991. I would like to sincerely thank all my helpers for their support and for providing many colour pictures from this period. My thanks go to – without military ranks, to avoid mistakes – Gerhard Albert, Karl Ascherl, Gerhard Ballhausen, Elmar Bauer, Helmut Baumann, Hans-Joachim Barakling, Günther Berg, Karl-Heinz Beuthe, Heinz Birkenbeil, Jürgen Bleil, Reinhart Blume, Ronald Bode, Helmut Borchers, Günter Borscheid, Peter Bündgen, Johann Büttgen, Jens Burkhart, Wolfgang Czaia, Dieter Dammjacob, Hans-Joachim du Roi, Hermann Eichin, Horst Fetzer, Klaus-Dietrich Flade, Klaus Forker, Gerd Gloystein, Wilhelm Göbel, Erhard Gödert, Camille Goossens, Georg-Wilhelm von Graevenitz, Günter Grondstein, Hub Groeneveld, Matthias Haas, Bruno Heyer, Josef Hoffmann, Klaus Homberg, Walter Jertz, Hartmut Jung, Jürgen Keuling, Burkhard Klesser, Peter Kmonitzek, Wolfgang Leuthner, Hans-Detlev Lichtenhof, Roger Lindsay, Hans-Peter Löffler, Rüdiger Lohse, Wolfram Lüders, Heribert Mennen, Horst Meyer, Hansachim Müller, Henning Müller-Nagell, Jürgen Neufeldt, Peter Nolde, Diether von Olleschik, Gert Overhoff, Axel Ostermann, Roland Pamler, Hubert Peitzmeier, Peter Petersen, Horst Philipp, Ralf-Dietmar Pilawa, Prof. Dr. Hans Pongratz, Helmut Predeschly, Henning Remmers, Achim Riedel, Horst Robitzkat, Jürgen-Wolfgang Rose, Hans-Dietert Rulle, Helmut Schaberl, Willy Scheungrab, Michael Schneider, Wilfried Schneider, Franz Schnell, Klaus Schönekäs, Gerhard Schöpke, Heinz Scholz, Joachim Schramm, Erwin Schroer, Gottfried Schwarz, Hans-Werner Schwiglewski, Klaus Seidel, Hubert Simon, Erika Söldner, Rudi Sonner, Peter von Stackelberg, Rainer Stadelmann, Dietrich Störmann, Joachim Streit, Rolf Stünkel, Ingomar Suhr, Joachim Thorwest, Heinrich Thüringer, Dieter Veit, Josef Voggenreiter, Klaus Wewetzer, Eckart Wienß, Wolfgang Wild and Berend Willigmann.

I would also like to thank Reinhard Dorn for the many picture edits. My sincere thanks go to the lectors Matthias Haas, Axel Ostermann and Henning Remmers and to Air Vice Marshal Sandy Hunter for his final proofreading of the English translations.

My heartfelt thanks go to Hans Jürgen 'Jack' Merkle, our Staffelkapitän – Commander – of the Cactus Starfighter Squadron and retired Generalmajor, for the foreword. His apt words show the bond and the fascination for our F-104 that we all enjoy to this day.

I also owe thanks to my dear wife, Elke, for her patience and support during the preparation for these books.

Finally, I ask for your forgiveness if I have omitted any of the many helpers.

Introduction

On 22 May 1991, a good 30 years ago, a Starfighter with a special white and blue livery took off from Ingolstadt-Manching for the last flight of a German Armed Forces F-104. That day marked the end of 30 years of use of this weapon system with the Iron Cross on its fuselage and wings. Even today, the aircraft exerts a great fascination for the observer with its shape, elegance and dynamics. For the pilots who were lucky enough to fly it, it was passion from the very first flight. For each of them, flying it was the fulfilment of a dream. But the technicians also bonded closely with their 104s. They were proud of their work and of their involvement with this combat aircraft, which was at the spearhead of NATO in the 1960s and '70s. Before the memory of this fascinating aircraft – the Starfighter – fades, I would like to breathe life into times long past with this colour-illustrated book about the history of the F-104 in the Luftwaffe, the German Air Force, and with the Marineflieger, the air arm of the German Navy. For all those who have flown, maintained or, in some other way, played a part in the operation of the Starfighter, these pictures will awaken memories of the years of Starfighter silhouettes in the skies over the Federal Republic of Germany.

Although I have made every effort to include previously unpublished images, this was not always successful, and it was occasionally necessary to break that rule, to avoid large gaps when looking back. Likewise, some of the colour images do not meet today's quality standards. Here, too, I have accepted slight deficiencies in some photos for historical reasons. My own nine years as a Starfighter pilot in Jagdbombergeschwader 31 'Boelcke' resulted in many pictures being offered to me by former members of that Wing. Some of the Jagdbombergeschwader (JaboG) 31 'B' pictures on the following pages are, therefore, representative of the events in the other German F-104 units. I dedicate this work to all the pilots who lost their lives while flying a German Starfighter. I personally knew many of those who died from 1972 onwards and some were close friends on the same squadron as me. They are not forgotten. I hope you will enjoy this collection of pictures, which is published in two volumes, and at the same time gain a revealing insight into this part of the Luftwaffe's history and the Marineflieger, now far in the past.

This volume covers the production, introduction and use of the Starfighter by the combat wings, while the second volume mainly covers the basic Starfighter training in the United States, weapon training in Europe, NATO competitions, accidents and some technical areas.

Klaus Kropf

Hauptmann Klaus Kropf in the cockpit of an F-104G before taking off for a training flight in autumn 1980. A few weeks later, he was on his way to RAF Cottesmore to start his Tornado conversion, followed by a three-year tour as a Tornado flying instructor. (Klaus Kropf)

Foreword

The last flight of a German Starfighter took place on 22 May 1991 at Test Centre 61 in Manching, officially ending the era of the most important flying weapon system of the German Air Force and Navy in the early years of the German Armed Forces. Klaus Kropf's illustrated book, with its great variety of pictures, provides a good opportunity to enjoy what we then experienced enthusiastically and uncritically, as well as to reflect now after the passing of time.

In addition, this meticulously compiled collection of photographs offers a sad but affectionate reminder of our deceased fellow pilots, especially the 116 comrades from the German Air Force, Navy and United States Air Force (USAF) who lost their lives in crashes. Even after many years, we have not forgotten them.

As the chairman of the Cactus Starfighter Squadron Association, the '104-spirit', which is still perceptible today but difficult for outsiders to grasp, became clear to me again when I looked at these images. Above all, a warm and comradely cohesion resulted from the joint training of Navy and Air Force pilots at Luke Air Force Base (AFB) in Arizona and continued in Germany within the operational units. Flying the 104 was a challenge, an effort and a pure joy at the same time. For us

Generalmajor Hans Jürgen 'Jack' Merkle in the cockpit of an F-104G Starfighter during his time as Officer Commanding (OC) Flying Group of Jagdbombergeschwader 34 at Memmingen, in summer 1984. (Hans Jürgen Merkle)

pilots, the myth of the Starfighter, which had already emerged at that time, was based on our sense of duty and unwavering respect in our dealings with our 'flying lady'.

The performance and historical classification of the approximately 1,800 Starfighter pilots trained in Germany – and especially in the US – goes far beyond their performance in the air. With the establishment of a very active foundation (International Friendship Foundation), we have kept alive the memory of the successful training at Luke AFB, we have given unlimited support to charitable institutions in the Phoenix area since 1979 and have promoted a German–American student exchange. All as a sign of gratitude to our American hosts and an expression of the German–American friendship. This objective is also served by an active partnership with an American flying squadron (F-16 training) at Luke AFB.

The Starfighter, the F-104G, significantly determined the combat value of the German Air Force and the German Naval Wings for more than 25 years and thereby shaped an era in military aviation. The '104' is thus also a symbol of the rebuilding of our armed forces after World War Two. And it represents our country's most important contribution as part of NATO in deterrence during the Cold War. We single-seat pilots played a major part in this with our high operational capability and readiness, and we can be proud of this. We did so for peace and freedom at that time. As the Chief of Staff of the German Air Force put it in the regulations governing Air Force Tradition, our performance and achievements 'are particularly appropriate when maintaining the traditions of the German Air Force'. The same is certainly true of our naval aviators!

With this in mind and with this well-illustrated book as a guide, we offer our own experiences and our understanding of 'Tradition' to the younger, active generations of military aviators.

Generalmajor Hans Jürgen 'Jack' Merkle

A backwards glance out of a TF-104G cockpit at a Jagdbombergeschwader (JaboG) 34 formation flying over snow-covered southern Germany. (Karl Ascherl)

Chapter 1

First Steps

The official formation of the German Armed Forces took place in November 1955, and the rapid build-up of several operational wings of the Luftwaffe (Air Force) and the Marineflieger (Naval Aviation) began immediately with the weapon systems F/RF-84F, Sabre VI, F- 86K and Sea Hawk. During this build-up phase, the Air Force Command had already begun the search for a successor model for these aircraft. To be an equal partner in the European spearhead of NATO in central Europe, the German Air Force had to be equipped with a modern aircraft of the latest generation at the beginning of the 1960s. For financial reasons, a successor model was sought that could take on all operational roles, from interceptor to naval fighter-bomber. In the course of 1957, three aircraft were finally selected. Besides the French Mirage III, two American models – the Grumman F11F-1F Super Tiger and the F-104 Starfighter – were considered. Test flights were conducted by Lieutenant Colonel Albert Werner and Major Walter Krupinski, Officer Commanding (OC) Waffenschule 30, the F-84F Weapon School. The small evaluation team proposed the selection of the F-104 as successor to most of the German Armed Forces' previous jet fighter aircraft. Lockheed promised a significantly improved Starfighter, which, after modifications to the airframe and the installation of a modern fire control system, a more powerful engine, an updated radar and a inertial navigation system, was to satisfy all the future tasks of the German Air Force. In autumn 1958, the Federal Ministry of Defence decided in favour of the F-104G, the 'all-rounder' Starfighter promised by Lockheed. The Defence Committee of the German

Above left: Clarence Leonard 'Kelly' Johnson, a long-time aircraft builder for Lockheed and the spiritual father of the Starfighter. Shortly before his death in December 1990, he met Wolfgang Czaia (see Volume Two: Epilogue) and presented him with this picture, with the dedication: 'Kelly Johnson – Best Regards to Wolfgang'. In his hands he is holding a model of the SR-71, which he developed as head of Lockheed's famous Advanced Development Programs department, also known as the 'Skunk Works'. (Wolfgang Czaia)

Above right: Starfighter badge.

First Steps

Parliament endorsed this decision and approved the purchase on 6 November 1958. Just a few months later, on 6 February 1959, the first purchase contract was signed with Lockheed for the acquisition of 66 F-104Gs. Barely six weeks later, on 18 March 1959, the purchase contract for 30 F-104F two-seaters and 210 F-104Gs, to be built under licence, followed. Less than a year later, in early February 1960, the first six Luftwaffe pilots flew to the United States to begin flight training. At the same time, the first training for technical personnel took place at Lockheed in Palmdale, California.

Thus began almost 30 years of employment of the Starfighter weapon system by the Luftwaffe and the Marineflieger.

```
                    MILITARY ASSISTANCE ADVISORY GROUP
                         FEDERAL REPUBLIC OF GERMANY
                         Box 810, APO 80 US FORCES

                                                     28 January 1960

SUBJECT:   Invitational Travel Order No 1-28 (AFS 117-60)
           Case Number:  GER 8128
                                                            Tng No
TO:        OBSTLT   LTCOL    GUENTHER RALL                 9425 GER
           HPTM     CAPT     HANS-ULRICH FLADE             9426 GER
           OLT      1STLT    BERTHOLD KLEMM                9427 GER
           OLT      1STLT    EDMUND ERNST SCHULTZ          9428 GER
           OLT      1STLT    WOLFGANG VON STUERMER         9429 GER
           LT       2NDLT    BERND KUEBART                 9430 GER

                              SECTION I
   1.  The Secretary of the Air Force of the United States authorizes and
invites you to proceed on or about 4 February 1960 from your present station
in the Federal Republic of Germany to Air Force Plant Representative
Lockheed Aircraft Corporation, Factory A-1  2555 N. Hollywood Way Burbank,
reporting not later than 6 February 1960 to attend:

   Course                                  Duration          Location
   Pilot Ground School/F-104D(F),          2 weeks           Burbank
   OCC1125 E-1.  Item 12522, FY 60.                          California
   Additional cross-training subsequent
   to OCC1125E-1:

   Pilot Transition/F-104D(F),             8 weeks           Palmdale
   OCC1125 E-2                                               California
```

Above: The travel order for the first six German pilots to fly to California in February 1960, under the command of Oberstleutnant Günther Rall for Starfighter training at Lockheed. The theoretical and practical training took place in the Conversion Flight over a period of ten weeks. (via Klaus-Dietrich Flade)

Right: After two weeks of theoretical training, a first look into the cockpit of an F-104F two-seater. (From left) Leutnant Bernd Kuebart, Lockheed test pilot Bob Faulkner and Oberleutnant Edmund Schultz. (via Klaus-Dietrich Flade)

Mid-February 1960 at Palmdale. One of the first Starfighter familiarisation flights is completed.
(via Klaus Kropf)

Oberstleutnant Günther Rall (right) and Oberleutnant Edmund Schultz after the start of the Starfighter training. On the morning of 24 February 1960, Rall was the first of the group to fly without an instructor. The other five course participants also flew their first solos on this day.
(via Klaus-Dietrich Flade)

Oberleutnant Wolfgang von Stürmer (right) before a flight with Lockheed test pilot Glenn 'Snake' Reaves. On the left is Hauptmann Hans-Ulrich Flade.
(via Klaus-Dietrich Flade)

Lockheed test pilot Bob Faulkner in front of the F-104F BB+363, serial no. 5050, shortly before the aircraft takes off for an extended training flight. Two more F-104Fs, with registrations BB+361, serial no. 5048, and BB+362, serial no. 5049, were available for the training at Palmdale. The F-104Fs were purely flying training aircraft, without a radar and a fire control system. (via Klaus-Dietrich Flade)

Intensive F-104G type trials took place in 1961 and 1962, conducted by the Joint Test Force at Edwards AFB, California. Divided into 'CAT I' and 'CAT II' test programmes, all subsystems such as the autopilot, the inertial navigation system and the fire control system were thoroughly examined for their required functionality during type testing. In addition, necessary improvements were specified. This image shows the pilots of the Joint Test Force at Edwards in spring 1962: (standing from left) Capitano Franco Bonazzi (Italy), Lieutenant Colonel Steve Cailleau (Belgium), Oberleutnant Erhard Gödert (Germany), Flight Lieutenant Bob Ayres (Canada), Captain Albert Crews (US), Kapitein Mathias Janssen (Netherlands); (kneeling from left) Major Heinz Birkenbeil (Germany), Major Norvin 'Bud' Evans (US). (via Erhard Gödert)

Two Lockheed-built F-104Gs in a formation flight during the Joint Test Force programme in spring 1962. (Airbus Corporate Heritage)

Left: Oberleutnant Erhard Gödert in front of a Luftwaffe F-104G at Edwards AFB in spring 1962. This Starfighter, built by Lockheed, with serial no. 2059 and the temporary registration KF+134, came to Germany in the summer of 1962 and was then flown by JaboG 31 'Boelcke', with the registration DA+106, later 20+51. (Erhard Gödert)

Below: This image was taken during a test flight as part of the Joint Test Force programme over the eastern part of California. This aircraft with serial no. 2016 was initially put into storage after arrival in Germany in autumn 1962. Following a later return to the United States, it was used for training German Starfighter pilots at Luke AFB, Arizona, from 1964 onwards. (Heinz Birkenbeil)

Chapter 2
Starfighter Production

The first purchase contract with Lockheed on 6 February 1959 was for the construction of 66 F-104G Starfighters at Lockheed in Palmdale, California. These F-104Gs were later dismantled after construction and brought to Germany by air or sea freight to be handed over to the Luftwaffe after reassembly by Messerschmitt (MTT) at Manching. Some of these F-104Gs initially remained at Edwards AFB for the Joint Test Force flights. The next contract of sale, dated 18 March 1959, was for 30 F-104F two-seaters to be built by Lockheed and for 210 F-104Gs to be built under licence by the Arbeitsgemeinschaft Süd (ARGE-South, Working Group South) in Germany. The ARGE-South comprised the companies Dornier, Siebel, Heinkel and MTT, with the final assembly of the aircraft taking place by MTT at Ingolstadt-Manching. Belgium and the Netherlands conducted evaluations of a new fighter aircraft in parallel with Germany. Here, too, the decision was made in favour of the F-104G, so that the three nations – Belgium, Germany and the Netherlands – agreed on a joint European Starfighter production. For this purpose, a co-ordination office, the NATO Starfighter Management Office (NASMO), was set up, with its headquarters at Koblenz, Germany. Italy also selected the F-104G as its new fighter aircraft in 1960 and joined the European Starfighter production. In addition to ARGE-South, three other production groups were formed to build the Starfighter in Europe. There was ARGE-North, with Hamburger Flugzeugbau, Weserflug, Aviolanda, Avio Diepen and Fokker, whereby the final assembly was carried out by Fokker at Schiphol, Netherlands. There was also ARGE-West, with Avions Fairy, FIAT, Siebel, Heinkel and SABCA. The final assembly of this group was handled by SABCA at Gosselies, Belgium. The third additional production group was ARGE-Italy, with Macchi, Piaggio, SACA, Aeronavali, SIAI-Marchetti and FIAT, with final assembly provided by FIAT at Turin-Caselle, Italy.

F-104G assembly in 1960 by Lockheed in Palmdale. (Lockheed)

Germany concluded further purchase contracts for 424 F-104Gs in 1960–61, with all but 30 to be built by European production groups. A total of 137 TF-104Gs were ordered between 1961 and 1965, again with the majority of the aircraft being built by European production groups under a complex work-sharing arrangement. In October 1969, a final contract was signed for the construction of a further 50 F-104G Starfighters to replace previous losses. These aircraft were built by ARGE-South and by VFW in 1971–72. The German Armed Forces ordered a total of 917 Starfighters. One F-104G from the ARGE- West was given to the Belgian Air Force as a replacement for a loss, so that a total of 916 Starfighters entered service with the German military.

In parallel to Starfighter production, the General Electric J 79-GE-11A engine was built under licence in Europe. Three companies in Belgium, Germany and Italy received the construction contracts, with each company producing about one-third of all engine components. The three companies were BMW-Triebwerksbau GmbH at Munich-Allach, Fabrique Nationale d`Armes de Guerres (FN) in Belgium and FIAT in Italy.

Registration and other markings being painted onto the fuselage at Lockheed in Palmdale. This F-104G, serial no. 2024, was airfreighted to Germany and, after reassembly at Manching in spring 1962, was handed over to JaboG 31 'B' with the registration DA+121. Just one year later, in spring 1963, the aircraft was again airfreighted to the United States and, after reassembly and necessary modifications, assigned to the German F-104 pilot training programme at Luke AFB, which began in 1964. On 16 September 1966, the aircraft crashed shortly after take-off at Luke AFB. The pilot, Captain James Torson (USAF), was fatally injured despite baling out from the aircraft. (Lockheed)

Starfighter Production

Right: Lined up at Palmdale in early 1961. The front three Starfighters are F-104As, modified and upgraded by Lockheed to F-104G/CF-104 standard. Four German F-104Gs are parked beyond. At the end of the line is a C-53 NASARR trainer, sold by Lockheed to the Luftwaffe a short time later (see Chapter 4: Jagdbombergeschwader (JaboG) – Fighter-Bomber Wings). (Lockheed)

Below: The ARGE-South (Southern Working Group) production. The supply chain with final assembly at Ingolstadt-Manching is shown. (Airbus Corporate Heritage)

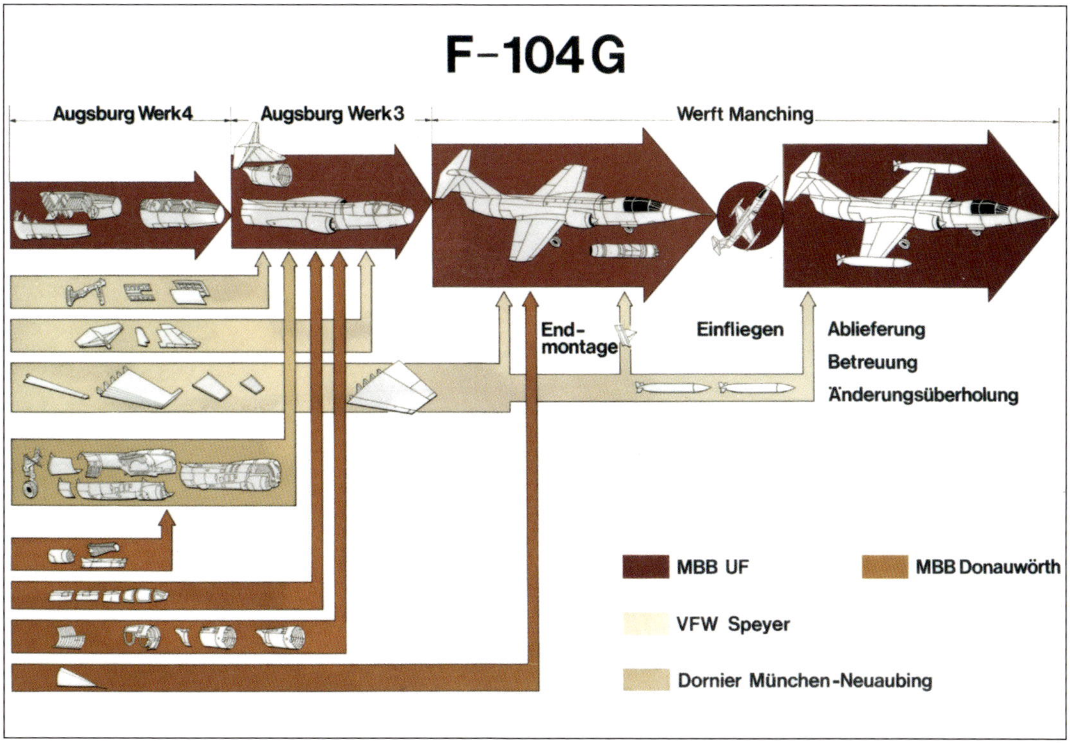

Above left: Starfighter badge.

Above right: Construction of the fuselage front section at Messerschmitt (MTT), later MBB, in Augsburg. (Airbus Corporate Heritage)

Left: F-104G final assembly at Ingolstadt-Manching in spring 1962. German licence-built and Lockheed-built F-104Gs undergoing assembly at Manching. (Airbus Corporate Heritage)

James Jester, chief pilot of MTT's Starfighter Production Test Flight Division, was the pilot of the first flight of the first F-104G assembled in Germany from components produced by Lockheed, on 10 August 1961. The aircraft had the serial no. 2009 and carried the registration DA+106. James Jester was greeted after the flight at Ingolstadt-Manching by representatives of the German Air Force and MTT. To the left of him, wearing a uniform cap, is Generalmajor Werner Panitzki, who became the second Luftwaffe Chief of Staff in October 1962. (James Jester)

The MTT-built F-104G with the temporary registration KE+439, serial no. 7139, on one of its first flights after completion in early 1964. The aircraft was handed over to JaboG 34 in April 1964 as DD+240, became 22+58 in 1968 and was flown by the Memmingen Wing until 1985. (Airbus Corporate Heritage)

Test flight from Schiphol, Netherlands, in the summer of 1964. This Fokker-built RF-104G with the temporary registration KG+325, serial no. 8225, was handed over to the Luftwaffe shortly afterwards. It received the Aufklärungsgeschwader (AG) 52 registration EB+123 and initially flew for about a year without camera equipment with Jagdgeschwader (JG) 71 'Richthofen'. After necessary technical work, it was assigned to AG 52 in the spring of 1967. It was only about four years later, with registration 24+76, that its use as a reconnaissance aircraft came to an end with the introduction of the RF-4E into the Luftwaffe (see Chapter 4). (Airbus Corporate Heritage)

Chapter 3
Waffenschule 10 (WaSLw 10) – Weapon School 10/Training Wing

Nörvenich Air Base, North Rhine-Westphalia, one squadron F-104F.
Moved to Jever Air Base, Lower Saxony 1964, built up to wing status.
F-104 in service 1961–83. Replaced by the Tornado.

Waffenschule (WaSLw) 10 was formed at Nörvenich in spring 1957 and moved to Oldenburg only a few months later, receiving the Sabre V. Its task was to train future day fighter pilots for the fighter wings, which were to be equipped with the Sabre VI weapon system. In April 1960, the newly formed 4. Staffel (4th Squadron) of WaSLw 10 began the training of F-104 pilots at Nörvenich. Initially, only the new F-104F two-seaters were available for this purpose. During 1963, the squadron received additional TF-104Gs, which allowed an expansion of the training programme.

Following the end of the Sabre VI training, in autumn 1963, the entire WaSLw 10 began its relocation, from Oldenburg and Nörvenich to Jever. This air base, located in East Friesland, had already been taken over from the Royal Air Force (RAF) at the end of 1961. It was then prepared for its planned use by the Starfighter, with extensive construction work including a runway extension.

On 7 February 1964, the Starfighter squadron flew its aircraft from Nörvenich to Jever, initially with 16 F-104Fs and TF-104Gs. The remaining aircraft and technical equipment followed a little later. Flight operations continued uninterrupted during the time of the relocation and acclimatisation at the new location.

Soon after arriving at Jever, WaSLw 10 began preparations for its new main task: 'Europeanisation'. This course was designed to familiarise all pilots returning to Germany from the Starfighter basic training course at Luke AFB, with European weather and airspace structure. In addition to theoretical training, the course comprised about 55 flying hours with many hours flown under instrument flight conditions and included flights to other European countries. Courses lasted about three months and were conducted until the last Starfighters were phased out from WaSLw 10 in 1983 and the Wing converted to the Tornado weapon system. In total, almost 1,000 young Air Force and Naval aviators, who had completed Starfighter training in the US, underwent Europeanisation.

In addition, more than 600 pilots coming from operational wings equipped with older aircraft, like the F-84F, Sabre VI or Sea Hawk, went through Starfighter training at the beginning of the 1960s. Other courses leading to qualification as an F-104 flying instructor or F-104 maintenance test pilot were part of the training syllabus. The Starfighters of WaSLw 10 flew a total of 123,728 flight hours between May 1960 and September 1983.

Above left: Emblem of Waffenschule (WaSLw) 10.

Above right: The first official Starfighter flight of the F-104 element of WaSLw 10 took place at Nörvenich on 22 July 1960. Lockheed's Chief Test Pilot, Bob Faulkner, flew with the Luftwaffe Chief of Staff, Generalleutnant Josef Kammhuber in the rear cockpit. The Secretary of Defence, Franz-Josef Strauß, greeted the General after landing. (via Klaus Kropf)

Pictured here, shortly after the arrival and reassembly of the F-104F in Germany in the summer of 1960, the BB+365, serial no. 5052, stands on a hangar apron at Nörvenich. The aircraft crashed on 19 June 1962 during a four-ship formation display practice, killing the pilot, Oberleutnant Bernd Kuebart. (Hans-Peter Löffler)

The engineers of 4. Staffel/WaSLw 10 prepare the F-104F BB+362, serial no. 5049, for a flight at Nörvenich in autumn 1960. The yellow engine ground-starting unit on the left, along with other devices of this type, came from USAF stocks and was replaced a few months later by ground-starting units, GPE-160, mounted on Mercedes-Benz Unimog vehicles. (via Klaus Kropf)

Above: The Luftwaffe's major airshow at Fürstenfeldbruck near Munich on 24 September 1961 was attended by some 500,000 visitors who watched displays by an F-104F four-ship formation in glorious weather. Six Starfighters were detached to Fürstenfeldbruck for this major event. (via Helmut Predeschly)

Below: With engines running, the pilots wait for the control tower's taxi clearance. A short time later, the formation takes off for a flight display over Fürstenfeldbruck. (Hans-Peter Löffler)

After the flight of the four-ship formation, some Starfighter pilots from WaSLw 10 were available to the press for questions and photos. From left: Oberleutnant Heinz Frye, Captain Bruce D Jones (USAF), Hauptmann Klaus Lehnert, Major Thomas Perfili (USAF), Oberleutnant Joachim Liedtke, Oberleutnant Hermann-Josef Sensen, Captain Jon Speer (USAF). (Hans-Peter Löffler)

A four-ship F-104F formation in spring 1962, during a training flight for a planned Starfighter display team. The first display of the team, including operational and aerobatic manoeuvres was scheduled for 20 June 1962, the day of the commissioning of JaboG 31 'B' with the Starfighter at Nörvenich. On 19 June 1962, during the last practice for the ceremonial event, the four-ship formation crashed east of Nörvenich. All four pilots were killed. Following this accident, no display team has ever been formed in the Luftwaffe (see Volume 2). (via Dieter Dammjacob)

During the Cuban Missile Crisis in October 1962, WaSLw 10 F-104Fs were loaded with Sidewinder AIM-9B air-to-air missiles and readied for possible use as interceptors. A practice Sidewinder is seen here on the wing tip of an F-104F at Nörvenich. (via Klaus Kropf)

In the spring of 1963, the F-104 component of WaSLw 10 at Nörvenich received the first TF-104Gs. In addition to flying training, these aircraft made possible weapon employment training fairly similar to the F-104G. Seen here in September 1963, next to a new TF-104G, are flying instructor Oberleutnant Hans-Joachim Barakling (left) with his Italian trainee, Tenente (Lieutenant) Antonio Stura (right). During the first years of Starfighter training, in addition to German pilots, many Belgian, Dutch and Italian pilots were trained. After his successful training, Tenente Antonio Stura remained as a flying instructor himself at WaSLw 10 for several years. (Hans-Joachim Barakling)

An F-104F during an approach to Nörvenich as part of the F-104 training programme for Luftwaffe and German Navy pilots in spring 1963. These pilots came from the F/RF-84F, F-86K, Sabre VI or Sea Hawk fighter aircraft. This F-104F has no arresting hook. With the Technical Instruction No. 50, the installation of arresting hooks on all German Starfighter aircraft was ordered in early 1963. (Hans-Joachim Barakling)

Two TF-104Gs of WaSLw 10 en route from Nörvenich on 12 October 1963 to the rendezvous point for a large formation flying over Wunstorf near Hanover as part of a 'field parade' to bid farewell to Federal Chancellor Dr Konrad Adenauer. In addition to transport aircraft and helicopters, the large formation comprised a total of 80 fighter aircraft, 16 of which were Starfighters. (Franz Schnell)

Several F-104Fs of WaSLw 10 in the north-west area of Nörvenich in autumn 1963. Two aircraft are already wearing the new camouflage paint. From this point onwards, all Luftwaffe and Navy Starfighters received the camouflage paint scheme of olive and basalt grey on the upper side and white aluminium on the underside of the aircraft. (Camille Goossens)

Above left: Taken after a training flight in autumn 1963, Capitaine Camille Goossens (right), a Belgian Starfighter flying instructor at WaSLw 10, is seen here, together with the Italian trainee Capitano Davide Albertazzi (left). (Camille Goossens)

Above right: Directly connected with the purchase of the F-104G Starfighter weapon system for the German Air Force was the F-102A Delta Dagger for the training of six German pilots at Perrin AFB, Texas, between September 1962 and March 1963. The F-104 was the second Luftwaffe combat aircraft after the F-86K to be equipped with an airborne radar. As the use of airborne radar in the interceptor version of the F-104G was considered demanding, the Luftwaffe High Command decided to give some young pilots a complete F-102A Delta Dagger training with the USAF before beginning their Starfighter training. Six young officers were selected to undergo the Undergraduate Pilot Training T-37/T-38 in the United States in 1961. The T-38 training took place at Webb AFB, Texas, and was the USAF's first T-38 course. This was followed by the F-102A Delta Dagger weapon system training, the Interceptor Pilot Training Course, at Perrin AFB with the 4780th Air Defence Wing. This training, involving about 85 flying hours, had as one focus the interception of air targets and the handling of the airborne radar. The six pilots were Leutnants Wolfgang Scheele, Hartwig Friedle, Gert Gollub, Horst Gebhard, Helmut Schaberl and Gert Fischer von Mollard. (Seen here, from left: Horst Gebhard, Gert Fischer von Mollard and Helmut Schaberl.) After returning to Germany in the spring of 1963, the six underwent F-104 training at Nörvenich and then started their operational service with JG 71 'R' at Wittmund. Three of them left the Luftwaffe only two years later, as their six years of service had come to an end. (Helmut Schaberl)

Several F-102s of the 4780th Air Defence Wing at Perrin AFB in early 1963. (Helmut Schaberl)

Evening engine test run in autumn 1963 on a parking area at Nörvenich. Noise protection devices were not yet in place. Justified noise complaints about the Starfighter flight operations lead to the relocation of the village of Oberbohlheim at the end of the 1960s. The small village lay directly outside the air base, in the western approach and take-off path. The new Oberbohlheim was built south of the air base. (Hans-Peter Löffler)

The relocation of the WaSLw 10 Starfighter training squadron from Nörvenich to Jever took place at the beginning of February 1964. A short time later, the Belgian flying instructor Capitaine Camille Goossens was photographed next to an F-104F at the new location. (Camille Goossens)

Above: WaSLw 10 at Jever took over the Europeanisation of young pilots after their Starfighter training at Luke AFB. Pictured here, the long line of Starfighter trainers, TF-104Gs and F-104Fs, at Jever in the summer of 1968. (Klaus Schöenekäs)

Right: In addition to the focus on instrument flying for familiarisation with European weather conditions, low-level navigation and formation flying with various flight manoeuvres were also part of the syllabus. The lead aircraft is seen here, climbing and pulling high G, from the 'close trail' position. (Jürgen Bleil)

The very first Starfighter built for the German Armed Forces in spring 1966 during a stopover at Nörvenich from Jever. This F-104F with registration BB+375, serial no. 5047, was built by Lockheed in 1959 and airfreighted to Technische Schule der Luftwaffe 1 (Technical School Air Force 1) at Kaufbeuren in early 1960. At the Technical School, the aircraft, which at that time had an F-104D canopy and the registration BF+011, was used for technical training until late spring 1961. Unlike the F-104F and TF-104G, the USAF's F-104D initially had no fixed centre section between the two left-opening canopies. After the installation of an F-104F canopy construction and various modifications, it was not until early 1965 that the aircraft was handed over to WaSLw 10 as BB+375. It replaced the F-104F, with the first registration BB+375, serial no. 5062, which on 29 March 1961 was the first Luftwaffe Starfighter to crash following engine failure. Hauptmann Hans-Ulrich Flade (OC 4. Staffel/Waffenschule 10) and Oberleutnant Wolfgang Strenkert managed to bale out. From 1968 onwards, the aircraft carried the registration 29+01. (Helmut Baumann)

TF-104G 28+23, serial no. 5953, in flight during May 1968. The aircraft was still equipped with C-2 ejection seats. The conversion of all German Armed Forces Starfighters to be fitted with the Martin-Baker GQ 7A ejection seats began in early 1968. A radioactive material warning symbol was visible on the rear fuselage section. From 1967, some TF-104Gs were equipped with a radioactive engine oil low-warning indicator. However, this was to little advantage. This modification was withdrawn after a short time and the affected aircraft reverted to the previous system. From 1974, the 28+23 flew with MFG 1, and later with MFG 2. (Dieter Veit)

Above left: Formation flight 'close trail'. (Dieter Veit)

Above right: A large formation with a total of 16 Starfighters of WaSLw 10 flying over a troop parade in May 1968. (Jürgen Bleil)

With a speed of 325 knots and flaps selected to 'Take-Off' position, this aircraft is approaching Jever. This TF-104G 27+01, serial no. 5701, was the first TF-104G built for the German Armed Forces by Lockheed. From 1963 to 1968, it carried the registration BB+101 and remained with WaSLw 10 until 1983. It was handed over to the Turkish Air Force in 1984. (Georg-Wilhelm von Graevenitz)

Flying hours anniversaries are always honoured in the Luftwaffe and among the Marineflieger. On 12 September 1974, a triple anniversary was celebrated at WaSLw 10 at Jever. Oberstleutnant Peter Treiber (left) with 2,000 Starfighter flying hours; Oberst Hans Klaffenbach (centre), OC WaSLw 10, with 3,000 total jet flying hours; and Hauptmann Rainer Stadelmann (right) with 1,000 Starfighter flying hours. Oberst Klaffenbach was OC WaSLw 10 for more than 11 years and is still remembered by many Starfighter pilots for his gruff manner. (Rainer Stadelmann)

In April 1971, a few weeks after the fatal crash of the F-104F 29+15, serial no. 5068, on approach to Jever AB following a boundary layer control duct failure, all remaining F-104Fs were decommissioned. Seen here, the F-104F 29+09, serial no. 5059 and former BB+372, during a stopover at Nörvenich at the end of 1970. (Helmut Baumann)

A WaSLw 10 formation of four over the East Friesland coast in the mid-1970s. (Rainer Stadelmann)

Low level near Helgoland, flying past the Lange Anna in the north-west of the island. (Ralf-Dietmar Pilawa)

The 20th Anniversary of WaSLw 10 was celebrated on 15 April 1977. The planned overflight of Jever by a large formation was successfully rehearsed a few days earlier. (Joachim Schramm)

Pitch-out or break – officially designated in the Luftwaffe as an airspeed reduction manoeuvre, seen here over Jever for a landing on Runway 28. (via Henning Remmers)

Above left: Two long-time Starfighter flying instructors of WaSLw 10 in January 1979 before a flight from Jever to Beja, Portugal. On the left is Oberstleutnant Horst Weidemann and, on the right, Oberstleutnant Hans-Joachim Barakling. (Hans-Joachim Barakling)

Above right: F-104G Starfighter of WaSLw 10 in a vertical climb. (Rainer Stadelmann)

A four-ship formation of WaSLw 10 consisting of two TF-104Gs and two F-104Gs over Jever towards the end of the 1970s. The picture certainly brings back memories for many pilots, as almost all of the German military Starfighter pilots took part in various F-104 training courses at Jever during their flying careers. (Hans-Joachim Barakling)

Chapter 4
Jagdbombergeschwader (JaboG) – Fighter-Bomber Wings

Jagdbombergeschwader 31 'Boelcke'
Nörvenich Air Base, North Rhine-Westphalia.
F-104 in service 1962–83. Replaced by the Tornado.

Jagdbombergeschwader 32
Lechfeld Air Base, Bavaria.
F-104 in service 1965–84. Replaced by the Tornado.

Jagdbombergeschwader 33
Büchel Air Base, Rhineland-Palatinate.
F-104 in service 1962–85. Replaced by the Tornado.

Jagdbombergeschwader 34
Memmingen Air Base, Bavaria.
F-104 in service 1964–87. Replaced by the Tornado.

Jagdbombergeschwader 36
Hopsten Air Base, North Rhine-Westphalia.
F-104 in service 1965–75. Replaced by the F-4F Phantom.

A total of 749 single-seat F-104Gs were built for the Air Force and Navy by Lockheed and the various joint ventures (ARGE) in Europe. They differed only slightly during construction. Towards the end of the individual aircraft production, decisions were made on their planned use and thus on necessary modifications or upgrades. For use in the fighter-bomber role, the number of upgrades was low because they were already incorporated in the fighter-bomber version during final assembly.

The Luftwaffe decided to convert five fighter-bomber wings equipped with the F-84F to the Starfighter, two in the 2 ATAF and three in the 4 ATAF. At the beginning of the 1960s, during the Cold War under NATO's 'Massive Retaliation' strategy, these wings were initially intended primarily for the nuclear role. The Luftwaffe's first Starfighter wing, Jagdbombergeschwader 31 'Boelcke' began its conversion in February 1962, and, on 20 June 1962, the unit officially entered service with the F-104G weapon system. The conversion of JaboG 33 began only slightly later in the summer of 1962, so that in spring 1963 one German Starfighter fighter-bomber wing was deployed in each of the 2 ATAF and the 4 ATAF areas and was working up to NATO assignment.

The three other F-84F wings, JaboG 32 at Lechfeld, JaboG 34 at Memmingen and JaboG 36 at Hopsten, received their Starfighters in the following years until 1965. Following NATO assignment,

'QRA' areas were introduced at the air bases concerned. These were high-security areas where several F- 104Gs loaded with US special (nuclear) weapons, initially B28 and, in later years, B43, B57 or B61, were on 15-minute standby. From the end of 1967, the new NATO strategy known as 'Flexible Response' came into force. This meant that not every attack against a NATO country was associated with a blanket nuclear response but that a flexible response could be made with conventional or individual nuclear weapons. For the F- 104G fighter-bomber wings, the new NATO strategy meant equal training for conventional and nuclear missions. Two German Starfighter wings, one each in the areas of the 2 ATAF and the 4 ATAF, JaboG 36 at Hopsten and JaboG 32 at Lechfeld, were completely withdrawn from nuclear assignment and reassigned to purely conventional NATO attack roles. For the pilots, the new NATO strategy resulted in a change of the Tactical Combat Training Programme (TCTP), namely in a significantly expanded training for conventional fighter-bomber roles to include army support (Close Air Support). JaboG 36 was the only F-104G fighter-bomber wing to be converted to the F-4F Phantom in the mid-1970s.

Until the retirement of the Starfighter in the 1980s, three of the four remaining F-104G fighter-bomber wings were operated in the dual nuclear/conventional role. After the conversion of the Luftwaffe's last F-104G wing, JaboG 34, to Tornado in autumn 1987, the Luftwaffe's Starfighter fighter-bomber wings had, together, flown a total of 973,805 flying hours between 1962 and 1987.

JaboG 31 'B'.   JaboG 32.   JaboG 34.

JaboG 33.   JaboG 36.

The first German operational wing, JaboG 31 'B' entered service on 20 June 1962 at Nörvenich with the F-104G. Only two months later, JaboG 33 at Büchel also received its first Starfighters. F-104G DC+102, serial no. 2069, of JaboG 33 following a landing at Nörvenich on a rainy day in autumn 1962. On the red ring behind the radar nose is the coat of arms of the 1. Staffel/JaboG 33. In 1967, this aircraft was airfreighted to the US and used for pilot training at Luke AFB until 1983. (Helmut Baumann)

In August 1962, JaboG 33 was the second wing of the Luftwaffe to begin conversion to the Starfighter. At the same time, the F-84Fs at Büchel maintained NATO assignment in the nuclear strike role for another two years. Due to the initially low number of allocated Starfighters combined with a low serviceability, some of the F-104-trained pilots of JaboG 33 occasionally flew the F-84F again. Although not recommended by the Luftwaffe High Command, such flights could not be completely avoided in order to maintain flying skills. This also applied to other Luftwaffe and Naval Air Wings during the conversion phase. A mixture of types can be seen on the May/June 1963 page of Hauptmann Gerd Gloystein's logbook, including flights in a T-33, F-84F, F-104G and Piaggio P 149. (Gerd Gloystein)

Seen through an F-84F cockpit windscreen are two F-104Gs of JaboG 31 'B' heading towards the rendezvous for a large formation with 102 fighter aircraft of the 2 ATAF on 26 June 1963. Oberleutnant Hans-Dietert Rulle of JaboG G 35 at Husum managed to take this photograph before the formation flew over the Headquarters 2 ATAF at Rheindahlen near Mönchengladbach. The occasion was the handover of command of the 2 ATAF from Air Marshal Sir John Grandy to Air Marshal Sir Ronald Lees. (Hans-Dietert Rulle)

The Luftwaffe's only C 53, the NASARR trainer with the registration XA+117, at Nörvenich in late summer 1962. The aircraft was purchased from Lockheed in 1961 and initially used as a radar instruction classroom for converting Starfighter pilots. NASARR training came to an end in 1964, and the aircraft was assigned to other tasks. (Helmut Baumann)

Gefreiter Helmut Baumann of JaboG 31 'B' is changing the nosewheel of DA+247, serial no. 2043, at Nörvenich in spring of 1963. This aircraft was used by JaboG 34 from 1966. However, it was briefly with JaboG 32 in spring 1984 with a white-blue special paint scheme for the last flight of the Lechfeld Wing. Back with JaboG 34, it was subsequently lost in a crash on 11 December 1984, killing the OC 2. Staffel/JaboG 34, Major Hans-Dieter Kerstan. (Helmut Baumann)

In the first years after the introduction of the Starfighter into the German Armed Forces, there was little hangar space to park the many aircraft. Keeping the aircraft outdoors had a negative effect on serviceability due to the effects of humidity and cold. From the mid-1960s onwards, additional hangarage was therefore built at all air bases. Seen here, a view of a ramp area at Nörvenich in the winter of 1962–63. (Helmut Baumann)

This F-104G DC+109, serial no. 7051, of JaboG 33 is shown here parked at Büchel on 24 March 1964 for maintenance work. Originally, all F-104Gs were handed over to the fighter-bomber wings with M 61 Gatling guns installed. Some of these aircraft subsequently received 'extended-range tanks' to increase their range for possible QRA missions. These additional tanks held about 800lb of fuel and were located at the front left of the fuselage instead of the M 61 Gatling gun. The cover on the gun port indicates this conversion. As early as summer 1965, this aircraft was handed over to JaboG 32 and remained there until the Wing was converted to the Tornado. In 1984, it was handed over to the Turkish Air Force. (Klaus Homberg)

Squadron exchange in March 1965. JaboG 31 'B' sent four F-104Gs to the 78th TFS 'The Bushmasters' of the USAF at RAF Woodbridge. This squadron was equipped with the F-101A/C. The pilots of the American squadron and the German detachment are seen here, standing in front of an F-101A and an F-104G. The Germans are (second from right) Oberleutnants Peter Thau, (fourth from right) Horst Völter, (seventh from right and OC 2. Staffel/JaboG 31 'B') Hauptmann Wolfgang Willam, (seventh from left) Leutnant Klaus Forker, and (third from left and technical officer) Oberleutnant Spiegelhauer. (via Helmut Baumann)

F-104G and an F-101A at RAF Woodbridge during the squadron exchange in March 1965. The engineering personnel of JaboG 31 'B' are on their way way to the F-101A for a short briefing. (via Helmut Baumann)

In February 1965, conversion began at Hopsten for JaboG 36, the Luftwaffe's last fighter-bomber unit to receive the Starfighter. A few weeks later, in spring 1965, two F-104Gs DF+124, serial no. 9150, and DF+117, serial no. 9134, are on Runway 01 before take-off at Hopsten AB. The emblem of JaboG 36 has not yet been applied. Both aircraft were handed over to the Luftwaffe a few months earlier by the Belgian aircraft manufacturer SABCA at Gosselies, Belgium. (Erwin Schroer)

JaboG 32 at Lechfeld received its first Starfighters in January 1965. Here is a view of the 'Toni' ramp in the south-east of the air base in the summer of the same year. The day's flight operations had ended. The engineering officer of the maintenance squadron, Oberleutnant Jürgen-Wolfgang Rose (centre), was on an inspection tour. With him were the engineers (from left) Oberfeldwebel Wiedemann, unknown, Stabsunteroffizier Böckler and Feldwebel Bennert. (Jürgen-Wolfgang Rose)

Above: Long standby duties at weekends often increase resourcefulness! Some engineers provided an example at Nörvenich in the mid-1960s. A Starfighter MB 7 drag parachute, jury-rigged to a Mercedes-Benz Unimog, brought the Unimog to a rapid stop after opening the parachute at full speed. (Helmut Baumann)

Left: A squadron exchange by JaboG 31 'B' took place in May 1966, in this case with the 55th TFS of the USAF at RAF Wethersfield. This squadron flew the F-100D Super Sabre. Welcomed by their American hosts, with a few cans of Budweiser beer, were (from left) Feldwebel Jürgen Thoms and Hauptmanns Andreas Kalkbrenner, Rüdiger Mazander, Franz Schnell and Hermann Hammerstein. (Elmar Bauer)

A German-Italian F-104G four-ship formation over northern Italy in the summer of 1967. The formation flight took place during a squadron exchange of 1. Staffel/JaboG 31 'B' with the 154° Gruppo of the Italian Air Force at Ghedi, Italy. (via Klaus Seidel)

Above left: In the summer and autumn of 1968, the runway at Nörvenich AB was extended. The flight operations of JaboG 31 'B' therefore took place from Büchel and Hopsten. Here is a view of several Starfighters at Hopsten in August 1968. F-104G 26+21, serial no. 9169, carries a BDU-9 concrete training bomb under the fuselage for special weapon B-28 delivery training. The F-104G to the left 22+38, serial no. 7116, had already converted to the Martin-Baker GQ 7A ejection seat and carries two underwing tanks with DayGlo paint. In March 1968, Technical Instruction 668 of the Luftwaffe Logistics Office ordered the application of DayGlo paint on the cylindrical parts of all Starfighter external tanks. Restricted to tip tanks a little later, all Luftwaffe and Navy Starfighters carried tip tanks with DayGlo paint until the 1980s. Both Starfighters shown here were taken out of service in the mid-1970s and used as war reserve (22+38) and scrapped due to corrosion (26+21). (Helmut Baumann)

Above right: On 1 January 1971, Generalleutnant Günther Rall became the Luftwaffe Chief of Staff. His predecessor, Generalleutnant Johannes Steinhoff, was bidden farewell on 16 December 1970 at Nörvenich during a large field parade. Front from left: Generalleutnant Johannes Steinhoff, Generalleutnant Günther Rall, General Ulrich de Maizière – German Armed Forces Joint Chief of Staff. (via Klaus Kropf)

Left: During the farewell ceremony for Generalleutnant Johannes Steinhoff on 16 December 1970, a 16-ship diamond formation of JaboG 31 'B', led by Major Hans-Joachim Barakling, OC 2. Staffel/JaboG 31 'B', overflies Nörvenich AB. (via Hans-Joachim Barakling)

Below: In spring 1971, this TF-104G 27+76, serial no. 5905, of JaboG 34 stands on Runway 24 at Memmingen prior to take-off. Until January 1968, the date of the changeover from individual wing registrations to the German Armed Forces registrations, the aircraft bore the JaboG 34 registration DD+379. With the exception of a few months in the entire period between 1965 and 1983, it was part of the JaboG 34 aircraft fleet. (Karl Ascherl)

With the introduction of the RF-4E Phantom weapon system for the two reconnaissance wings – AG 51 'I' and AG 52 – in 1971, the RF-104Gs were converted into fighter-bomber and fighter versions and distributed accordingly. This RF-104G 24+76, serial no. 8225, was parked at Nörvenich in spring 1971, just after a brief handover of the aircraft from AG 52 to JaboG 36 and the AG 52 wing crest had not yet been removed. At the end of 1971, the aircraft was converted to the fighter version and assigned to JG 71 'R'. (Helmut Baumann)

Shown here after take-off, at Nörvenich in autumn 1971, the rotation of the main landing-gear legs of the F-104G 22+10, serial no. 7080, can be seen during retraction. This was necessary to accommodate the two main wheels in the slim fuselage. (Walter Jertz)

Above: Several F-104Gs of JaboG 36 on the ramp in the north-west of Hopsten AB in the early 1970s. At this time, hardened aircraft shelters (HAS) had not yet been built and aircraft were parked in rows for flight operations at many air bases. (Günter Grondstein)

Right: Flying low over northern Germany. The effectiveness of the disruptive camouflage paint scheme is clearly evident. (Walter Jertz)

Left: Six pilots of JaboG 34 at Memmingen each reached 1,000 flying hours on the Starfighter on 13 April 1972. Seen here, after reporting to the Commander 1. Air Division, Generalmajor Carl-Heinz Greve, and awaiting congratulations: (from left) Major Jürgen Stehli (OC 2. Staffel/JaboG 34), Hauptmanns Rolf Reinert, Herbert Hesse, Oberleutnant Gerhard Schöpke, Hauptmann Adolf Maier and Oberleutnant Rolf Gensheimer. (via Gerhard Schöpke)

Below: An open day at Büchel AB on 2 September 1973. A mixed formation consisting of a Mirage III of the French Air Force and two F-104Gs of JaboG 33 are breaking formation prior to landing. (Klaus Homberg)

Oberstleutnant Anton Weilnböck, OC Flying Group JaboG 36 at Hopsten, reached the end of his tour in September 1972. On this occasion, a farewell photo was taken with most of the Wing's pilots: (seated front row from left) Oberleutnant Volker Neuenfeldt, Hauptmann Hartmut Gernhuber, Oberleutnant Manfred Hopf, Hauptmann Wolfgang Ralser; (standing from left) Oberleutnants Klaus Mildenberger, Wolfgang Kuhlen, Karl Freese, Oberfeldwebel Peter Esther, Oberleutnant Günter Hein, Oberfeldwebel Harald Schumann, Oberleutnants Johannes-Dieter Hassenewert, Wolfgang Süßschlaf, Klaus-Dieter Girnus, Hermann Brose, Oberstleutnant Anton Weilnböck, Oberstabsarzt 'Doc' Karl-Heinz Schneider, Hauptmann Hans Rusche, Oberleutnant Kurt Simonis, Hauptmann Detlev Schulte-Bisping, Oberstleutnants Dieter Stephan, Peter Treiber (OC 2. Staffel/JaboG 36), Major Joachim Hoppe, Oberleutnants Gerd Schmidt, Heinz Knöllingerl; (sitting on wing from left) Oberfeldwebel Josef Pauli, Hauptmann Rolf-Dieter Walbeck, Major Dirk Ortmann, Oberleutnant Hans-Richard Beyer; (sitting on fuselage from left) Hauptmann Hans-Jörg Nebel, Oberleutnant Meinhardt Feuersenger, Hauptmanns Peter Speetzen, John Miller, Oberleutnant Jürgen Becke. (via Hubert Peitzmeier)

Above left: Low-level flight at high speed involves risks, including bird strikes. During a squadron exchange of JaboG 31 'B' with the 112th Filo of the Turkish Air Force at Murted AB near Ankara in early September 1973, low-level flights were carried out for navigation training in Turkish airspace. During one of these flights 21+63, serial no. 7032, flew through a flock of seagulls, some of which penetrated the canopy and subsequently hit the helmet of the pilot, Hauptmann Volker Ding. In addition, the ejection seat was severely damaged and the upper seat handle was partially pulled out. Fortunately, the engine was not damaged and Hauptmann Ding, who was only slightly injured, was able to land the Starfighter safely at Murted. (Helmut Baumann)

Above right: Hauptmann Volker Ding's Gentex HGU-2A/P pilot helmet with destroyed sun visor after the bird strike over Turkey in early September 1973. (Helmut Baumann)

Above: Eight Starfighters of JaboG 31 'B' seen on their way to Runway 25 at Nörvenich on 21 May 1974 – two separate four-ship formations that took off a short time later for weapons training on air-to-ground firing ranges. The aircraft were carrying Mk 25 practice bomb carriers with DM 18 practice bombs under the fuselage. (Günter Grondstein)

Right: During the period of the transition to the F-4F Phantom of the Hopsten Wing in 1975, its Starfighters often took off after stopovers at Nörvenich with a cow emblem instead of the Wing's Westphalian prancing horse on the fuselage. Thanks to some F-104 engineers, it was a humorous expression of healthy rivalry between the two wings. (Heribert Mennen)

Runway construction work forced the wings concerned to temporarily relocate their flight operations to other air bases. For this reason, elements of JaboG 31 'B' moved to Erding near Munich for several weeks in autumn 1975. Seen here are two F-104Gs in front of an aircraft shelter at the Erding AB shortly before starting their engines for a low-level flight over southern Bavaria. Next to the Volkswagen Bulli in the foreground is a GPE-160 mounted on a Mercedes-Benz Unimog, supplying electrical power and compressed air to start the J-79 engine. (Klaus Kropf)

Above left: The six engine instruments to the right of the two artificial horizons during a flight at high level. The nozzle indicator at the bottom left shows the correct, almost completely closed position of the nozzle for this flight condition. In the first years of Starfighter operations, engine malfunctions such as 'open nozzle' were the cause of several crashes. Technical modifications to the emergency nozzle closing systems were carried out. (Klaus Kropf)

Above right: A four-ship of JaboG 34 Starfighters in close formation south of Memmingen AB. (Traditionsgemeinschaft JaboG 34 Allgäu)

A formation landing at Getafe, Spain, in February 1976. This F-104G 23+48, serial no. 8027, of the JaboG 31 'B' and flown by Hauptmann Rüdiger Lohse, is just touching down on Runway 05. Bluish smoke from the rubber abrasion of the landing-gear tyres can be seen just behind the aircraft. The Spanish AB Getafe, near Madrid, was a popular destination for the crews of many Luftwaffe and German Navy combat aircraft undertaking navigation training flights in the European NATO area. There were hardly any pilot and weapon system officers of the 1970s and '80s – whether they flew F-104s, G.91s, F/RF-4s or Tornados – who did not get to know the Bodega del Norte in the old town of Madrid. This is where they met to enjoy Spanish brandy together, and also to fill one or two plastic canisters with the brandy to enjoy later, at home. (Klaus Kropf)

The pilots of the 1. Staffel/JaboG 31 'B' in December 1976: (kneeling from left) Hauptmanns Arno Ewald, Franz Franzl, Helmut Wörner, Oberleutnants Dieter Knittel, Helmut Gattke, Hauptmann Harald Oelkers, Oberleutnants Dieter Erbe, Peter Kornmann, Hauptmann Siegfried Kahler, Oberleutnant Bodo Burmann; (standing from left) Hauptmanns Joachim Gottschalk, Wolfgang Baltes, Volker Ding, Erwin Schaupp, Jörg Stock, Oberleutnants Lothar Chwallek, Karlfried Rasch, Hauptmanns Wolfgang Grünert, Hans-Josef Niesen, Oberleutnant Volker Ebeling, Major Peter Pyczak (OC), Oberleutnants Günther Blecks, Aarne Kreuzinger-Janik, Gerald Backermann, Hauptmann Herbert Kaiser, Oberleutnants Hans-Günther Glanz, Hans-Peter Giezek; (at the cockpit from left) Hauptmanns Rolf-Joachim Kohlhoff, Wolfgang Halbey. (via Klaus Kropf)

After a weekend in Spain in February 1979, Hauptmann Rüdiger Lohse (left) and Hauptmann Matthias 'Mattes' Haas preparing TF-104G 27+07, serial no. 5708, for its return flight to Nörvenich. The aircraft was in service with JaboG 31 'B' from 1964 to 1981 and was subsequently handed over to the Greek Air Force. The helmet placed on the front windscreen frame is an SOS Gueneau 316. These French helmets replaced the American Gentex helmets in the Starfighter, Phantom and G.91 squadrons in the mid-1970s. They were slightly heavier than Gentex helmets and had an integrated clear visor and sun visor. (See also the SOS Gueneau helmet in the picture in the Foreword, page 7.) (Klaus Kropf)

During a squadron exchange with the 102° Gruppo of the Italian Air Force at Rimini, Italy, a two-ship formation F-104G of JaboG 31 'B' and F-104S of the 102° Gruppo flies over the south-eastern part of Emilia Romagna in August 1977. As was often the case during a squadron exchange, the German technicians had affixed the emblem of their own wing to the host aircraft. (Klaus Kropf)

During the Cold War, many European air bases were prepared to receive US operational units, with appropriate logistics and infrastructure. In the 1970s, Nörvenich AB also became a Co-located Operating Base (COB). In September 1977, part of the 301st TFW, stationed at Carswell, Texas, and equipped with F-105Ds, deployed to Nörvenich for training and to test all necessary preparations. This picture, shot by the author from the rear cockpit during an F-105F flight, was taken shortly before the return to Nörvenich from a co-ordinated mission with both the F-104 and the F-105. (Klaus Kropf)

A TF-104G of JaboG 32 shortly after take-off from Runway 21 of Lechfeld in autumn 1977. The white, long building of Luftwaffe Technical School II can be seen in the background. The TF-104G 27+38, serial no. 5740, was flown by JaboG 32 from the end of 1966 until 1984. (Hans Pongratz)

Regular passenger flights by the wing's flight surgeons were very welcome. Stabsarzt Dr. Hans Pongratz (left), JaboG 32's flight surgeon for several years, is pictured here on 5 January 1978 after his farewell flight before leaving the Wing. His pilot on that day in the TF- 104G 27+36, serial no. 5738, was Hauptmann Karl-Heinz Brandner (centre). Hauptmann Axel Bree (right) was also there to say goodbye to his former flight surgeon. This aircraft was one of the six TF-104Gs handed over to the Italian Air Force in the mid-1980s. (Hans Pongratz)

Above left: On 22 May 1978, 2. Staffel/JaboG 34, the Edelweiss Squadron, was twinned with No. 17 Squadron RAF. This squadron was based at RAF Brüggen, west of Düsseldorf, at the time and was equipped with the Jaguar GR1 fighter-bomber. Seen here, a mixed formation of Jaguar GR1s and F-104Gs over the foothills of the Alps. (Airbus Corporate Heritage)

Above right: Major Eugen 'Ike' Gisder after his last Starfighter flight in spring 1980 at Nörvenich. With his direct and brash manner, 'Ike' is remembered by many Starfighter pilots, as he could look back on assignments as a T-38 flying instructor in the USA, F-104 flying instructor with WaSLw 10 at Jever and weapons instructor with the Luftwaffe Weapons Training Center at Decimomannu, Italy. His last Starfighter flight took place during a short attachment to JaboG 31 'B'. Major 'Ike' Gisder (left), is welcomed here by Hauptmann Dieter Erbe (centre) and Major Walter Jertz (OC 1. Staffel/JaboG 31 'B') (right). (via Walter Jertz)

Left: During a deployment of JaboG 31 'B' to Erding in summer 1978, Hauptmann Helmut Gattke achieved 1,000 total flying hours. He received congratulations from wing staff officers after the landing: (from left) Oberstleutnant Alexander Heuser (OC Flying Group), Oberstleutnant Gert Overhoff (Deputy OC Wing) and Hauptmann Walter Jertz. (via Hartmut Jung)

Below: The colourful badge 'TIGER BY THE TAIL', for more than 1,000 flying hours with J 79 engines, adorned many a flight jacket. See above right, Major Eugen 'Ike' Gisder.

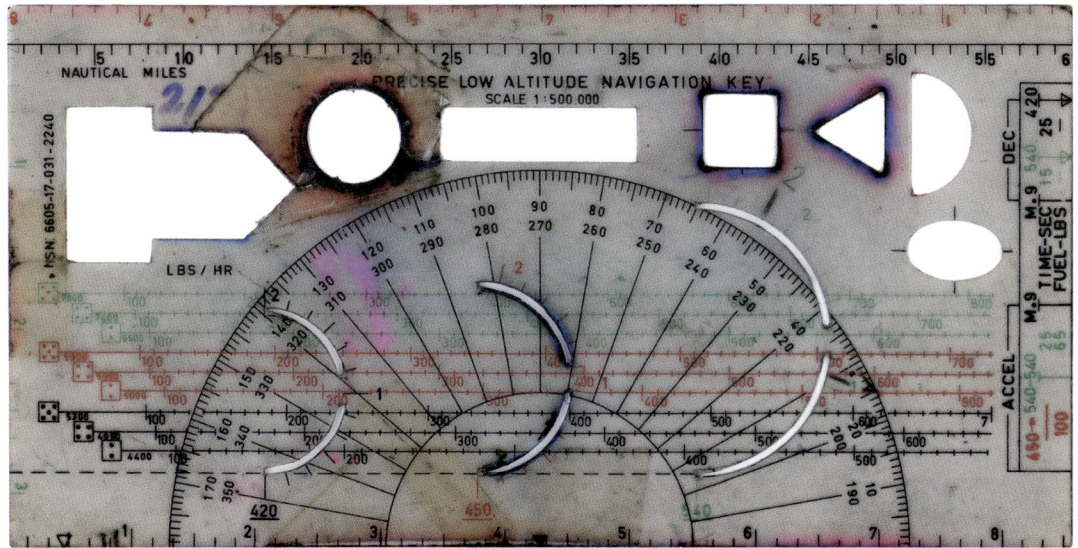

The Litton tool, named 'Precise Low-Altitude Navigation Key'. An indispensable aid for planning Starfighter low-altitude navigation in the speed range 420 to 540 knots above ground. After nine years of use by the author, signs of wear and tear are clearly visible. Good flight and chart preparation, as well as a high level of attention during the flight, are imperative. Visual low-level navigation requires constant monitoring of the required course and the timed route while, at the same time, checking the navigation chart on a scale of 1:500,000, which was often attached to the pilot's kneeboard. Added to that was manual control of the aircraft at a height of between 250 and 1,000ft above ground and checking the planned ground speed. The Starfighter was even used for Close Air Support with a Forward Air Controller (FAC) during army exercises. Determining target co-ordinates transmitted by radio from the FAC on the 1:50,000 scale map used for this purpose while flying very low, identifying the target and bringing the aircraft into the correct position for a simulated weapons release required the utmost concentration. In today's world of GPS, moving maps and computer-aided navigation and weapon systems, this seems impossible. But in the Starfighter, this was the normal workload of the pilots of the fighter-bomber squadrons. It was not until 1970–71 that groundspeed indicators and radar altimeter displays were installed in the German Starfighter cockpits. (Klaus Kropf)

View of a page in a low-level navigation booklet. Produced for the Starfighter fighter-bomber squadrons by the Radar Prediction Centre at Büchel, these pre-prepared booklets showed the radar image to be expected for all sections of the route. The radar predictions appeared next to the general navigation information on each page and the booklets were used for radar low-level navigation training by day and night. This illustrated page shows the approach from the north to the air-to-ground firing range at Nordhorn. (Horst Meyer)

Memmingen in July 1980. Five pilots of JaboG 34 made their last Starfighter flight. After landing, they could look back on a total of 20,256 flying hours in the Luftwaffe. All five left the Air Force shortly afterwards and retired: (from left) Hauptmanns Rolf Gensheimer, Horst Hein, Werner Weiss, Oberstleutnant Dietrich Hartmann, Major Paul Signus. (via Karl Ascherl)

Squadron exchange of 2. Staffel/JaboG 32 with the 192nd Filo of the Turkish Air Force at Murted, Turkey, in the summer of 1980. The participating pilots during the visit to Turkey: (rear from left) Hauptmanns Rainer Seifert (Detachment Commander), Wolfgang Pfizenmaier, Oberleutnant Klaus-Peter Uhler; (front from left) Hauptmanns Wilhelm Tekath, Reiner Mößner, Wolfgang Leuthner. The F-104Gs flown to Turkey as part of the squadron exchange all bore the 2. Staffel/JaboG 32 squadron emblem. (Wolfgang Leuthner)

F-104G 21+68, serial no. 7037, of JaboG 33 in early summer 1981 over the Eifel hills. (Gottfried Schwarz)

The pilots and other personnel of 1. Staffel/JaboG 34 in December 1980 at Memmingen: (kneeling from left) Oberleutnant Volker Zirfaß, Hauptmanns Thomas Würzner, Burghard Frick, Günther Borscheid, Oberleutnant Albrecht Braig, Hauptfeldwebel Gerd Siebert (engineering warrant officer), Oberleutnants Peter Hartmann (concealed behind Walter Quintel), Hauptfeldwebel Herbert Vogg (Do-28 pilot), Horst Menzel (crew bus driver); (standing from left) Hauptmanns Heinrich Schneider, Andris Freutel, Wolfgang Thoelke, Major Karl Ascherl, rear unknown, Oberleutnant Jürgen Gundling, rear unknown, Hauptmanns Eberhard Menrad, rear unknown, Helmut Plum, Oberleutnants Lutz Mühe, Werner Rast, Major Eckhard Sowada (OC), Oberleutnants Rainer Baldus, Wolfgang Zuber, Captain Charly Grant (USAF), Hauptmann Ingo Balszun, Hauptmann Manfred Molitor, Erwin Jäger (crew bus driver); (on wing) Hauptmann Berend Reefmann; (dressed in white, kitchen staff) Conny Bentele and Erich Klas. (Traditionsgemeinschaft JaboG 34 Allgäu)

A two-ship formation of JaboG 32 waiting for take-off clearance in September 1981 at Lechfeld. The low-level route was planned to enter an air-to-ground firing range to drop high-drag practice bombs DM 18. Two of these small blue practice bombs can be seen on the Mk 25 bomb carrier under the fuselage. The correct adjustment for take-off of the horizontal stabilisers is also clearly visible – aligned with diagonal white stripe on the fin. (Günter Grondstein)

A four-ship F-104G formation of JaboG 33 standing on the runway at Büchel in spring 1982. The practice bombs under the Mk 25 practice bomb carriers and the LAU 10 rocket launchers under the left wings indicate that the flight will soon be on its way to an air-to-ground firing range. (Gottfried Schwarz)

A two-ship F-104G formation of JaboG 31 'B' at Nörvenich seen shortly after take-off in autumn 1982. The aircraft 20+50, serial no. 2058, had been visiting RAF Bentwaters shortly before as part of a squadron exchange with the 78th TFS 'The Bushmasters' of the USAF. Since 1979, this squadron had been equipped with the A 10 Thunderbolt and its squadron crest is painted on the Starfighter fuselage behind the engine intake. The lead aircraft, 23+87, serial no. 8094, flew for several years as RF-104G with AG 51 'I' and then was converted to a fighter-bomber after introduction of the RF-4E Phantom into the Luftwaffe. It was assigned to JaboG 31 'B' in 1971. (Heribert Mennen)

Above left: Flight hour anniversaries are also celebrated in the Eifel. In September 1982, Hauptmann Joachim Riedel (right) of JaboG 33 at Büchel achieved 2,000 hours total flying time. The squadron mates of 1. Staffel/JaboG 33 greeted him warmly and presented him with a bouquet of dried meadow flowers. On the left is Hauptmann Lothar Schmidt. (via Achim Riedel)

Above right: The Commander Luftwaffe Operations Command, Generalleutnant Fritz Wegner, undertook his last Starfighter flight on 1 March 1983 in a TF-104G of JaboG 32, together with Oberstleutnant Werner Sadrina, OC Flying Group JaboG 32. After landing at Lechfeld, Generalleutnant Wegner (right) was greeted by many of his commanders: (from left) Generalmajor Fred Noack (Commander 1. Air Division), Oberstleutnant Helmut Borchers (Deputy OC JaboG 34), Oberst Fritz Morgenstern (OC JaboG 32). (Military Historical Collection Lechfeld)

Hohenzollern Castle near Hechingen in Baden-Württemberg, the ancestral seat of the Prussian royal family, is located on the hilltop of Hohenzollern and is a well-known navigational reference point. A Büchel F-104G two-ship, in close formation, is seen here in front of the castle in early summer 1983. (Gottfried Schwarz)

Jagdbombergeschwader (JaboG) – Fighter-Bomber Wings

The OCs of all combat aircraft wings, or their deputies, at Lechfeld after the last Starfighter flight of Generalleutnant Fritz Wegner on 1 March 1983: (from left) Oberstleutnant Manfred Michen (JaboG 43), Oberst Manfred Purucker (AG 51 'I'), Oberst Georg Müller (JaboG 35), Oberst Botho Engelien (JaboG 36), Oberstleutnant Josef Wagner (JaboG 44, reserve unit, not active), Oberst Lothar Kompch (JaboG 71 'R'), Oberst Fritz Morgenstern (JaboG 32), Oberst Hans-Heinrich Block (JaboG 74 'M'), Oberst Albert Weber (JaboG 49), Generalleutnant Fritz Wegner, Oberst Johannes Glowka (JaboG 33), Oberstleutnant Helmut Borchers (JaboG 34), Oberstleutnant Wilhelm Göbel (WaKo), Oberst Gert Overhoff (JaboG 31 'B'), Generalmajor Fred Noack (1. Air Division), Oberst Jürgen Schlüter (JaboG 41), Oberst Siegfried Thormann (AG 52), Oberstleutnant Hans-Peter Koch (Deputy OC JaboG 32). (Military Historical Collection Lechfeld)

The very first F-104G of the Luftwaffe 20+01, serial no. 2001. The first flight of this Starfighter took place in October 1960 at Lockheed in Palmdale, California. Following transport to Germany and handover to the Luftwaffe, the aircraft was always operated by JaboG 31 'B'. Seen here after landing on 30 April 1983, on an open day at Nörvenich, when the aircraft flew with the Wing for the last time. Soon after, the aircraft was handed over to the technical unit Luftwaffenschleuse 11 at Ingolstadt-Manching. Requests to keep this F-104G in Germany for historical reasons were not followed up by the Luftwaffe High Command. Fregattenkapitän Wolf-Dietrich Havenstein of the Erprobungsstelle 61 therefore flew 20+01 to Murted, Turkey, in January 1985 for handover to the Turkish Air Force. (Günter Grondstein)

The first Starfighter of the Luftwaffe to bear a colourful special paint scheme was 28+31, serial no. 5961, of JaboG 31 'B'. In April 1983, it was painted in blue with the large white sword of the Boelcke coat of arms on both sides of the fuselage to mark the end of flight operations with the Starfighter and the simultaneous celebration of the Wing's 25th anniversary. Flying the 28+31 in this picture in May 1983 was Oberleutnant Reinhard Helbig with air traffic control officer Oberleutnant Heribert Mennen in the rear cockpit. (Heribert Mennen)

At the beginning of 1983, 28+31 received its special paint scheme. What is done with foils nowadays still had to be achieved with conventional paint in the 1980s. Stabsunteroffizier Harry Biernat is seen here, spraying the blue paint on the fuselage. (Heribert Mennen)

There is always a dedicated team behind a 'special painting' project. The blue 28+31 of JaboG 31 'B' was the work of five of the Wing's servicemen. After completion in April 1983, OC JaboG 31 'B' congratulated them before the first flight with the special livery: (from left) Oberst Gert Overhoff, Leutnant Gereon Nellessen, Stabsunteroffizier Peter Will, Oberleutnant Heribert Mennen, Oberleutnant Reinhard Helbig. Not pictured is Stabsunteroffizier Harry Biernat. (Heribert Mennen)

Twenty-five years of JaboG 32 at Lechfeld. The tail unit of 26+30, serial no. 9182, received a Bavarian livery to mark the occasion in July 1983. (Wolfgang Leuthner)

Above left: A TF-104G of JaboG 34 in the summer of 1983 approaching the Spanish AB Getafe, near Madrid. The runway can be seen below the aircraft. In the background is the sea of houses of the Spanish capital. (Ralf-Dietmar Pilawa)

Above right: Father and son in a TF-104G of JaboG 34 in summer 1983. Generalmajor Hans-Ulrich Flade, one of the first German Starfighter pilots and now Chief of Staff Headquarters 2 ATAF, together with Oberleutnant Klaus-Dietrich Flade. He had just completed training in the USA and was posted to JaboG 32. Klaus-Dietrich Flade later became a test pilot and was known to the public as a cosmonaut of the MIR '92 mission in March 1992. (Klaus-Dietrich Flade)

In September 1983, about six months after the start of flight operations with the Tornado by JaboG 31 'B', a detachment of the 154° Gruppo from Ghedi, Italy, was welcomed as a guest at Nörvenich. The Italian squadron was also being converted from the Starfighter to the Tornado at the time. Part of this formation was a TF-104G of JaboG 32 from Lechfeld. Deliveries of Tornados to this Starfighter wing began in 1984. (Heribert Mennen)

In March 1984, this F-104G 23+32, serial no. 8007, of JaboG 32 landed at Nörvenich on Runway 07, deploying its BS-4000 drag parachute. Many years earlier, in 1965 and 1966, the aircraft had been used for SATS tests in the USA in Lakehurst, New Jersey. After additional SATS trials in 1966 at Lechfeld – with the registration DB+257, the end of this project and the removal of the SATS equipment – the aircraft was assigned to JaboG 32 (see Volume Two). About 18 months after this landing at Nörvenich, on 3 October 1985, and by then on the inventory of JaboG 34 at Memmingen, the aircraft crashed after an engine failure over the Suippes air-to-ground firing range in eastern France. The pilot, Major Aarne Kreuzinger-Janik, OC 1. Staffel/JaboG 34, managed to bale out and later, as a Generalleutnant, became the 14th Luftwaffe Chief of Staff. (Heribert Mennen)

Lechfeld in April 1984. A few days before the end of Starfighter operations by JaboG 32, eight F-104Gs taxi to the runway at Lechfeld. After take-off, they formed the Wing's last large F-104 formation before Tornado operations began. (Military Historical Collection Lechfeld)

F-104G 20+36, serial no. 2043, in April 1984 with the white-blue special livery for the Starfighter Farewell by JaboG 32. (Military Historical Collection Lechfeld)

Above: F-104G 24+19, serial no. 8161, of JaboG 34 received the temporary registration 25+50 and a black-red-gold special paint scheme for the celebration on 5 May 1984. A large coat of arms of JaboG 34 adorned the tail unit. (via Hans Jürgen Merkle)

Right: On 5 May 1984, JaboG 34 in Memmingen celebrated 25 years as a fighter-bomber wing and 50 years of the Memmingen AB with an open day. A special letter imprint was produced for this occasion. (Helmut Borchers)

Below: During the JaboG 34 open day on 5 May 1984, a 'Le Mans launch', with a total of 16 F-104Gs was demonstrated. The leader was Oberstleutnant Hans Jürgen 'Jack' Merkle, OC Flying Group. He took off with the special black-red-gold livery F-104G. The many visitors at the Memmingen AB reacted enthusiastically to the impressive take-off spectacle. (via Jürgen Neufeldt)

The large formation of 16 F-104Gs seen overflying Memmingen AB on 5 May 1984, with its landing gear extended and landing lights on. (Helmut Borchers)

Above: A two-ship formation of JaboG 33 in the spring of 1985. The lead aircraft was already wearing the new dark all-round camouflage paint. Both aircraft were later allocated to JaboG 34. (Ronald Bode)

Left: After 30 May 1985, only one Luftwaffe operational wing, JaboG 34 at Memmingen, was still equipped with the Starfighter. This picture shows the Wing's F-104G 22+62, serial no. 7143, as part of a two-ship formation in a steep high-G climbing turn. (Traditionsgemeinschaft JaboG 34 Allgäu)

At the end of the JaboG 33 flight operations with the Starfighter at Büchel, F-104G 21+67, serial no. 7036, appeared in a yellow and black special paint scheme, with the coat of arms of Rhineland-Palatinate on the tail. The pilot of the last flight, on 30 May 1985, was Oberstleutnant Jürgen Stehli, the Deputy OC. He is seen here standing in front of 21+67, which was handed over to the base technical training workshop after the flight. Today it is still on display at the main gate of Büchel AB. (JaboG 33)

In October 1986, F-104G 21+91, serial no. 7060, of JaboG 34, on approach to Nörvenich. Six months later, it was taken out of service and, in spring 1989, transported by rail to the Ailes Anciennes museum in Toulouse, France. Several aviation museums in Germany and Europe received a F-104 as an exhibit after the end of the German Armed Forces' Starfighter flight operations. (Helmut Baumann)

Starfighters as far as the eye can see! In May 1987, many F-104Gs were to be seen at Ingolstadt-Manching awaiting future employment, such as transfer to Greece or Turkey. They are seen here, lined up on the northern runway of the air base – a beautiful sight near the end of the long service life of the F-104 Starfighter weapon system in the Luftwaffe and with the Naval Air Wings. Altogether in the 1980s, the Turkish Air Force received 165 F-104Gs and 36 TF-104Gs from the Bundeswehr's inventory. The Greek Air Force received 42 F-104Gs, 16 RF-104Gs (formerly of the German Navy) and 23 TF-104Gs. (via Klaus Kropf)

On 23 October 1987, JaboG 34 at Memmingen also mounted its last Starfighter flight with great sadness. The picture shows a formation of 16 F-104Gs during a training flight for the planned overflight of Memmingen AB on 23 October. Unfortunately, it was cancelled for weather reasons. (Gottfried Schwarz)

F-104G 22+55, serial no. 7135, with its special white and blue livery is the farewell aircraft of the Memmingen Wing. On 23 October 1987, the Deputy OC, Oberstleutnant Edgar Fischer, flew 22+55 for the last time during the event, marking the end of the F-104 and the beginning of Tornado operations. The aircraft spent all its life with JaboG 34 from June 1964 to October 1987. After the last flight, it was handed over to the technical training workshop of the base. (Ronald Bode)

JaboG 34's 'farewell aircraft' with its special white and blue livery in flight. (Gottfried Schwarz)

Chapter 5
Jagdgeschwader (JG) – Fighter Wings

> **Jagdgeschwader 71 'Richthofen'**
> Wittmund Air Base, Lower Saxony.
> F-104 in service 1964–74. Replaced by the F-4F Phantom.
>
> **Jagdgeschwader 74 'Mölders'**
> Neuburg/Donau Air Base, Bavaria.
> F-104 in service 1963–74. Replaced by the F-4F Phantom.

Two Luftwaffe fighter wings, the Sabre VI-equipped Jagdgeschwader (JG) 71 'Richthofen' at Wittmund and the F-86K-equipped JG 74 at Neuburg/Donau, converted to the Starfighter respectively in 1963 and 1964. The F-104 impressed with powerful acceleration and the ability to reach twice the speed of sound. On the other hand, it quickly became clear that a tight turning radius was not achievable in air combat. Pilots spoilt by victorious 'dog fights', primarily with the Sabre VI, had to adapt quickly and mourned their previous aircraft a little. Without external tanks, missions at high altitude of up to one hour were possible, but in low-level flight, two external tanks had to be carried to achieve this flight time. For the fighter mission, two Sidewinder AIM 9B air-to-air missiles were loaded in addition to the M 61 Gatling gun. The Aero 3B Catamaran fuselage carriers, introduced for this purpose in 1969, were only used for a short time, as they had a negative effect on flight characteristics in certain load configurations. This resulted in the most common configuration of two underwing tanks and the air-to-air missiles at the wingtips for low-altitude interceptions.

In the early years, the F 15A NASARR radar required the pilots to make many inputs on the radar control panel when attacking an air target in the air-to-air mode. Through extensive radar modifications towards the end of the 1960s, as well as an improvement of the infrared seeker of the air-to-air missile AIM 9B, the effectiveness of the Starfighter in the interceptor role was greatly enhanced. A secondary air-to-ground role was evaluated towards the end of the 1960s. The so-called 'Tactical Fighter' capability was, however, not realised; only air-to-ground gunnery remained part of the training.

Nevertheless, the time of the F-104G weapon system at the two fighter wings was coming to an end after a little more than ten years. The Luftwaffe had decided to introduce the F-4F Phantom. It carried a greater weapons load, had a significantly smaller turn radius in air combat and had a two-man crew, with a division of labour between pilot and weapon system officer. In 1974, the Starfighters of the fighter wings were decommissioned and most of them scrapped in the following years. This was despite the fact that only a few of these aircraft had flown more than 1,500 hours. The additional construction of 50 F-104Gs in 1971–72 made them surplus to requirement.

JG 71 'R'.   JG 74.

In late spring 1963, a few weeks after the arrival of the first Starfighters on JG 71 'R' at Wittmund, F-104G JA+101, serial no. 8010, is pictured in flight. The aircraft was not the first F-104G of the Wittmund Fighter Wing. It came from Belgian production at SABCA at Gosselies and was handed over to JG 71 'R' at the beginning of May 1963. By April of that year, some Starfighters from the SABCA and Fokker production lines had already arrived at Wittmund. (Gerhard Albert)

JA+116, serial no. 8029, shortly after handover to JG 71 'R' in July 1963, high over the North Sea coastline. (Diether von Olleschik)

Several Starfighters on the apron in front of 2. Staffel/JG 71 'R' in the north-east of Wittmund AB in summer 1963, just a few months after the arrival of the first F-104Gs on the Wing. JA+237, serial no. 8054, and later on 23+64, was built by Fokker, Netherlands, while JA+248, serial no. 9012, and later on 25+66, was built by SABCA, Belgium. This aircraft initially carried the wing emblem on the vertical stabiliser. However, in the end, the engine air intake proved to be the best location for the emblem of the proud Luftwaffe Richthofen Wing. (Diether von Olleschik)

Above: F-104G JA+104, serial no. 8015, was one of the few Starfighters not handed over directly from the manufacturer to one of the two fighter wings. It flew for four months in early 1963 as DA+115/DA+124 with JaboG 31 'B' before it was transferred to JG 71 'R' in July 1963 and decommissioned in May 1974. (via Henning Remmers)

Right: A four-ship F-104G formation of JG 71 'R' seen from below in the spring of 1965. The launchers for Sidewinder AIM 9B air-to-air missiles can be seen on the wingtips. (Gerhard Albert)

The same formation as in the previous picture. All four aircraft had received the disruptive camouflage paint scheme in 1964. (Gerhard Albert)

Above: Starfighter JD+243, serial no. 6618, of JG 74 during take-off at Neuburg/Donau in June 1965. The aircraft was built the year before by FIAT in Turin, Italy. As with all German F/RF-104Gs built by FIAT, the registration and Iron Cross are positioned too high on the forward fuselage. (BArch, Image-F027409-0005/Photo: Berretty)

Left: Neuburg/Donau in June 1965 at JG 74. Hauptmann Josef Herbst is seen here, checking the aircraft's logbook before a flight in JD+117, serial no. 9117. (BArch, Image-F027404-0003/Photo: Berretty)

Above left: A picture for the press. Twelve F-104Gs of JG 74, with pilots, lined up at the beginning of Runway 27 at Neuburg/Donau. Hauptmann Peter Cramer can be seen in the first Starfighter, JD+107, and Oberleutnant Lutz Moldenhauer in the second, JD+250. The other pilots can not be identified. The registration of JD+107, serial no. 9067, was changed to 25+88 in 1968. (See other 25+88 images in this chapter.) (BArch, Image-F027407-0008/Photo: Berretty)

Above right: Another press photograph – a four-ship F-104G formation of JG 74 over the foothills of the Alps. (BArch, Image-F027411-0006/Photo: Berretty)

Night flying at Wittmund. JA+250, serial no. 9014, on its way to the runway. (Peter Nolde)

In July 1967, 2. Staffel/JG 71 'R' carried out a squadron exchange with 2 Squadron of the French 13 Fighter Wing – 02/013 'Alpes' – stationed at Colmar, France. This squadron was equipped with the Mirage IIIE. Here, a mixed formation of F-104Gs and Mirage IIIEs is seen flying during the squadron exchange. Hauptmann Wilhelm Göbel was piloting the lead aircraft, with Oberleutnant Peter Vogler in the 'slot'; Capitaine Huerre was flying the Mirage IIIE on the left, and Sergent Cavalin was piloting the Mirage IIIE on the right. Parts of the East Frisian Islands are visible in the background. (Peter Nolde)

Above left: Christmas party at 2. Staffel/JG 71 'R' in December 1966 at Wittmund. Like everywhere else in the Luftwaffe, all ranks meet at the end of the year for a cosy get-together with coffee and cake: (from left) Gefreiter Rudi Grötzner, Stabsunteroffizier Peter Nolde, Hauptmann Gerhard Albert, Hauptfeldwebel Erich Pankonin, unknown. (Peter Nolde)

Above right: Mid-1968 at Wittmund. Hauptfeldwebel Rudi Hennig of JG 71 'R' is seen preparing the cockpit for a training flight. He wears a NORGE 3F life jacket under his BA 15 parachute. These life jackets, already introduced at the time of the first-generation jets such as the Sabre and F-84F, were only used until the end of the 1960s. In parallel, SECUMAR 10F lifejackets were introduced at the beginning of the 1960s. A trigger handle can be seen at the top of the C-2 ejection seat. This was used to disconnect the steel cables of the foot restraining system in case of an emergency on the ground, necessitating the rescue of the pilot. (Peter Nolde)

A F-104G from JG 71 'R' 26+01, serial no. 9123, seen in summer 1968 at high altitude over north Germany. The aircraft carries wingtip-mounted Sidewinder air-to-air AIM 9B missiles. A programme to replace the Lockheed C-2 ejection seat by the Martin-Baker GQ 7A seat in the entire German Starfighter fleet began at the beginning of 1968 but had not yet been incorporated in this aircraft. (Wilhelm Göbel)

Quick Reaction Alert (QRA) duty at Neuburg/Donau in hot July 1970. QRA duties required pilots of the two fighter wings at Wittmund and Neuburg/Donau to take-off within five minutes of an alert and then to follow the CRC's target instructions. Seen here, Hauptmann Georg-Wilhelm von Graevenitz enjoys the sunshine during one of his QRA duties. Boots and flight suit at the ready, his F-104G is in the background, prepared for an immediate engine start. (Georg-Wilhelm von Graevenitz)

A formation take-off by a pair of F-104Gs of JG 71 'R' in autumn 1970 on Runway 26 in Wittmund. Both aircraft are 'clean', without external tanks. (Peter Nolde)

Above: Beautiful aerial view of a two-ship F-104G formation of JG 74 in August 1971. (Horst Fetzer)

Right: Back at Neuburg/Donau in January 1972 after a flight from RAF Wattisham. As part of a squadron exchange by 1. Staffel/JG 74 with No. 29 Squadron RAF flying Lightning F3s a number of pilots visited Wattisham for several days: (standing from left) Hauptmanns Jürgen 'Keule' Schubert, Alfons 'Whiskey' Wieshuber, Oberleutnant Franz Schatt, Oberleutnant Stuwe (technical officer), Major Wilhelm Göbel (OC 1. Staffel/JG 74); (kneeling) Oberleutnant Werner Vogin. (via JG 74)

In close formation with JG 74 F-104G 21+06, serial no. 6628, in August 1971. (Horst Fetzer)

Four TF-104G of JG 74 in the parking area for the two-seaters at Neuburg/Donau in August 1971. (Horst Fetzer)

After a stopover at Nörvenich at the end of May 1973, this F-104G of JG 74 25+93, serial no. 9081, is seen on its way to Runway 07. It is armed with two air-to-air AIM 9B (mod) Sidewinder missiles. These AIM 9Bs had an improved infrared seeker, recognisable by a darker nose cap. The stopover was necessary after providing an escort for the Ilyushin 62 of the General Secretary of the Central Committee of the KPDSU, Leonid Brezhnev, who was coming to Bonn for a state visit to the Federal Republic of Germany. (Helmut Baumann)

In September 1973, Lightning F3s of No. 5 Squadron from RAF Binbrook were guests of 1. Staffel/JG 74 at Neuburg/Donau. Seen here, a mixed formation consisting of two F-104Gs from the Neuburg Wing, led by a Lightning F3 of No. 5 Squadron, flies high over southern Germany. (via JG 74)

Several F-104Gs of the two fighter wings, JG 71 'R' and JG 74, deployed to Hopsten in late summer 1973 as part of a NATO exercise. For interceptions at low altitudes, a configuration with two underwing tanks was usually chosen, allowing AIM 9B air-to-air missiles to be carried on the wingtips and giving a low-level flight time of over an hour. In 1969, Aero 3B Catamaran fuselage air-to-air missile carriers were introduced on both fighter wings. Used almost exclusively by JG 74, they were, however, taken out of service again after about two years. (Günter Grondstein)

TF-104G 27+41, serial no. 5743, of JG 74, shortly before landing at Nörvenich in autumn 1973. During the conversion of the Neuburg Fighter Wing to the F-4F Phantom in the summer of 1974, this Starfighter was initially assigned to JaboG 34. In 1984, like five other TF-104Gs in the mid-1980s, it was handed over to the Italian Air Force. It was then flown by the Starfighter Training Squadron, 20° Gruppo at Grosseto, Tuscany, until 1997. (Heribert Mennen)

Starting in 1974, the Starfighter was replaced by the F-4F Phantom in the two fighter wings JG 71 'R' at Wittmund and JG 74 'M' at Neuburg/Donau, as well as in the fighter-bomber wing, JaboG 36, at Hopsten. The Neuburg Wing had been granted the honorary title 'Mölders' on 22 November 1973, but this was removed by parliamentary decree in March 2005. After the arrival of the first F-4F at Wittmund, a detachment from JG 71 'R' deployed to Nörvenich with some of its F-104Gs. This allowed several pilots waiting for Phantom conversion training to maintain their flying status. F-104G 23+53, serial no. 8032, is seen here parked at Nörvenich in early 1974. The aircraft was transferred to LwVersRgt 1 at Erding in May of that year, used for spare parts and subsequently scrapped. (Heribert Mennen)

A sad sight in the summer of 1975. Many Starfighters formerly flown by JG 71 'R' and JG 74 'M' are seen in a fenced-off area at Erding. After spare parts had been removed, dismantling and scrapping began. F-104G 25+88, in the middle, can be seen on other images in this chapter during its years of service with JG 74. (Klaus Kropf)

Chapter 6
Aufklärungsgeschwader (AG) – Reconnaissance Wings

Aufklärungsgeschwader 51 'Immelmann'
Ingolstadt-Manching, Bavaria.
The wing moved to Bremgarten Air Base, Baden-Württemberg, in 1969.
F-104 in service 1963–71. Replaced by the RF-4E Phantom.

Aufklärungsgeschwader 52
Leck Air Base, Schleswig-Holstein.
F-104 in service 1964–71. Replaced by the RF-4E Phantom.

The two reconnaissance wings, Aufklärungsgeschwader (AG) 51 'Immelmann' at Ingolstadt-Manching, Bavaria, and AG 52 at Leck, Schleswig-Holstein, received their first RF-104Gs in 1963 and 1964, respectively. Like all other RF-104Gs of the German Armed Forces, the aircraft were built by ARGE-Nord, with final assembly at Fokker at Schiphol, Netherlands, or by ARGE-Italy, with final assembly at FIAT in Turin. For the reconnaissance mission, the optical camera system TA-7M from the Dutch manufacturer De Oude Delft was installed in the lower fuselage behind the nosewheel.

Due to the limited space available in the fuselage, the system consisted of three cameras with a short focal length (2 x 52mm and 1 x 70mm). This was the only way to cover a terrain section of 120° below the aircraft. The cameras used 70mm film which, after landing, was prepared for evaluation using normal photographic processing.

'Visual tactical air reconnaissance with photo confirmation' was the operational task of the reconnaissance wings. This meant that the pilots had to find and identify possible targets in low-level flight at high speed while running the cameras. Ideally, possible targets were overflown. An 'in-flight report' was given by radio during the return flight back to base. After landing, a detailed debriefing of the pilot took place, with the film developed by specialist personnel accompanying the debriefing and serving as confirmation of the target descriptions. From the point of view of photo reconnaissance, the RF-104G's camera system was a significant step backwards compared to the previous RF-84F camera system, which had up to six long-focal-length cameras. Moreover, the RF-104G, like its predecessor, remained a fair-weather reconnaissance aircraft. It was, therefore, not surprising that the Luftwaffe soon looked to replace the Starfighter in the reconnaissance role.

The decision was made in favour of the RF-4E Phantom, and the introduction of the new weapon system began in 1971. Two years earlier, AG 51 'I' had already relocated to Bremgarten, south-west of Freiburg. The pilots of the two reconnaissance wings went into the RF-4E conversion training with very mixed feelings. They were looking forward to a very modern reconnaissance system and, at the same time, were bidding a sad farewell to a fantastic aircraft.

The RF-104Gs released by the introduction of the RF-4E were converted for further use in fighter-bomber and fighter aircraft.

Above left: AG 51 'I'.

Above right: AG 52.

Left: RF-104G 24+82, serial No. 8232, of AG 52 during an evening flight in June 1968 over Schleswig-Holstein. The three De Oude Delft TA7M cameras are located behind the small bulge at the front of the underside of the fuselage. (Joachim Streit)

AG 51 'I' received its first Starfighters in late autumn 1963, including RF-104G EA+105, serial no. 8095, built in spring 1963 by Fokker in the Netherlands. The aircraft is parked at the beginning of 1965 in the northern area of Ingolstadt-Manching – the Wing's base until spring 1969. (Heinz Scholz)

RF-104G EA+252, serial no. 6627, of AG 51 'I' in the summer of 1966 near Ingolstadt-Manching. The aircraft, built by FIAT in Italy, crashed on 8 June 1967 after an engine failure on approach to its home base. Oberfeldwebel Reinhold Mast managed to bale out safely. (Heinz Scholz)

At Ingolstadt-Manching in spring 1966, Oberleutnants Rudolf Sonner (left) and Dietmar Treppe are about to take off on low-level flights, tasked to provide 'visual reconnaissance with photo confirmation'. A few weeks later, on 13 June 1966, Oberleutnant Treppe was one of the two pilots killed when his RF-104G of AG 51 'I' collided with another, over the Netherlands. The other pilot killed in the collision was Oberleutnant Manfred Hippel. (Rudolf Sonner)

In the 1960s, many pilots, as well as a large number of technical personnel, decided to stay in the German Armed Forces for a limited number of years only. Here we see Hauptmanns Jürgen Schott (left) and Rudolf Sonner (right) towards the end of their service in March 1967, together with Major Gerhard John, OC 1. Staffel/AG 51 'I'. Jürgen Schott and Rudolf Sonner continued their flying careers with major airlines in Germany and Switzerland. (Rudolf Sonner)

EA+122, serial no. 8164, of AG 51 'I' in early 1967 after landing and taxiing back to the ramp at Ingolstadt-Manching. In 1968 the aircraft was given the registration 24+22, and in 1971 it was converted into a fighter-bomber. In the following years, it was flown almost exclusively by JaboG 31 'B'. (Rudolf Sonner)

Formation take-off of two RF-104Gs of AG 51 'I' in early 1967 at Ingolstadt-Manching on Runway 25L. The large hangar of Erprobungsstelle 61 – Test Centre 61 – is in the background. (Airbus Corporate Heritage)

AG 52's open day at Leck on 1 June 1967. In sunny weather, a four-ship RF-104G formation flies over the air base, with landing gear extended. (via Willy Scheungrab)

Also on display during the open day at Leck on 1 June 1967, along with many other aircraft, was this TF-104G EB+373, serial no. 5920, of AG 52. Newly built in spring 1966, the Starfighter was assigned to AG 52 and received the new registration 27+90 in 1968. After the wing's conversion to the RF-4E Phantom in 1971, it remained for many years with the German Weapons Training Center in Sardinia, Italy, with few interruptions. (via Willy Scheungrab)

24+01, serial no. 8132, of AG 51 'I' during a low-level flight over southern Germany in early 1968. The change from the C-2 to the Martin-Baker GQ 7A ejection seat has not yet taken place. The wing emblem of AG 51 'I', the flying owl, adorns the nose of the aircraft. (Heinz Scholz)

Shortly after an evening take-off at Leck in June 1968. The RF-104G of AG 52, here flown by Hauptmann Lutz Mundhenk, shines in the glow of the setting sun. The Fokker-built aircraft 24+82, serial no. 8232, was converted to the fighter-bomber version at the end of 1970 and then assigned to JaboG 31 'B'. Barely eight weeks later, on 30 April 1971, the aircraft crashed after an engine failure during a low-level flight over northern Germany. The pilot, Hauptfeldwebel Stefan Liebold, managed to bale out safely. (Joachim Streit)

In February 1970, several RF-104Gs of the AG 52 parked on the ramp in the south-eastern part of Leck AB for daytime flight operations. On the right is a hangar built in the mid-1960s for night storage of the aircraft. (Klaus Kropf)

Taxiing after landing at Hopsten AB in the early summer of 1968, TF-104G 27+35, serial no. 5737, of AG 52 was transferred to JG 71 'R' at the end of 1971 and moved to JaboG 31 'B' at the beginning of 1974. The Starfighter wings all had several TF-104Gs for advanced training and annual instrument flight tests for the pilots. In 1984, the 27+35 was handed over to the Turkish Air Force. (Klaus Kropf)

RF-104 23+92, serial no. 8102 of AG 51 'I' during a high-altitude flight over Norway in October 1970 with Hauptmann Arthur Hanawitsch in the cockpit. Only a short time later, the conversion of the two reconnaissance wings to the RF-4E Phantom began in January 1971 at Bremgarten and in August 1971 at Leck, respectively. (Joachim Streit)

Aufklärungsgeschwader (AG) – Reconnaissance Wings

The end of the RF-104G weapon system's service life with the two Luftwaffe reconnaissance wings was near. 24+60, serial no. 8208, was assigned to AG 52 at the end of September 1970 for the last few months, coming from AG 51 'I'. The previous wing emblem, the AG 51 'I' flying owl on the front fuselage, was immediately removed in a makeshift manner – shown by a dark spot at the front under the cockpit. For AG 52, with its heraldic panther, the flying owl on the fuselage was not an acceptable emblem! This RF-104G was converted back to the fighter-bomber version at the end of 1971 and subsequently assigned to JaboG 34. (via Willy Scheungrab)

In June 1971, a detachment of AG 52 took part in the Tiger Meet for the last time with the Starfighter. At the same time, the host USAF air base, Upper Heyford, UK, held an open day. Here, Hauptmann Armin Falk is seen standing by the pitot tube of his RF-104G 24+70, serial no. 8219, answering questions from the visitors. A few months later, the aircraft was converted to the fighter version, assigned to JG 74 and decommissioned as early as 1975. (MAP Aircraft Photographs)

Nice view of the underside of an RF-104G in flight. The camera windows of the De Oude Delft cameras are clearly visible under the front fuselage. (Joachim Streit)

Chapter 7
Marinefliegergeschwader (MFG) – Naval Air Wings

Marinefliegergeschwader 1
Schleswig-Jagel Naval Air Station, Schleswig-Holstein
F-104 in service 1963–81. Replaced by the Tornado.

Marinefliegergeschwader 2
Eggebek Naval Air Station, Schleswig-Holstein
F-104 in service 1965–86. Replaced by the Tornado.

The two Naval Air Wings Marinefliegergeschwader (MFG) 1 at Schleswig-Jagel and MFG 2 at Eggebek, both in Schleswig-Holstein, began converting to the Starfighter MFG 1 in autumn 1963 and MFG 2 in spring 1965. The Marineflieger Command had recommended the British Blackburn Buccaneer, a heavy fighter-bomber in service with the Royal Navy, as a replacement for the Sea Hawk reconnaissance and fighter-bomber aircraft that was flown up to then. However, for financial reasons, this special path could not be followed by the naval aviators. It was only possible to achieve simpler, better and more cost-effective logistics, infrastructure, training and personnel arrangements through the procurement of a very large number of F-104 aircraft.

The naval aviators' area of operation was the North Sea and the Baltic Sea, and operations against sea targets in these areas were their main task. The pilots accepted the F-104G with pleasure. It was very stable in low-level flight over the sea, had excellent acceleration and a small radar cross section. The engine was very reliable and well able to cope with the conditions of flight over sea. The armament initially consisted of the onboard gun, ballistic high-explosive bombs and unguided rockets. In the 1970s, these were supplemented by cluster bombs, AS 30 air-to-ship guided missiles and the air-to-ship 'fire and forget' homing missile, Kormoran 1. AIM 9B air-to-air missiles were also available to engage aerial targets.

The 27 RF-104G reconnaissance aircraft used by the 1. Staffel/MFG 2 were, like the RF-104Gs of the Luftwaffe, equipped with the TA-7M camera system of the Dutch manufacturer De Oude Delft. The naval aviators also suffered the shortcomings of this system until an improved reconnaissance system was installed from December 1978 onwards.

MFG 1 was the first German Armed Forces wing to be converted to the Tornado from the end of 1981. MFG 2, on the other hand, continued to fly the Starfighter with the black anchor on the fuselage for several years. It was not until May 1987 that the last Marine Starfighter flight took place during a large farewell event at Erding.

MFG 1.

MFG 2.

In September 1963, MFG 1 at Schleswig-Jagel received its first Starfighters. It was not until over a year later, in March 1965, that the first Starfighters for MFG 2 arrived at Eggebek. This formation of six Marineflieger F-104Gs is seen here in June 1965. The lead aircraft was from MFG 2, VB+245 with serial no. 7202. It had only been handed over to the Wing at the beginning of the month, newly built. The other five aircraft were from MFG 1. With the exception of VA+103, serial no. 7083, all Starfighters wore the standard disruptive camouflage paint scheme. VA+103 received this camouflage during an overhaul in spring 1966. (BArch, image-F027434-0011/photographer: Berretty)

During a Marineflieger exercise in mid-June 1965, several pilots of MFG 1 and MFG 2 wait in front of 1. Staffel at Schleswig-Jagel for their next mission: (back row from left) Kapitänleutnants Knut Winkler, Helmut Kröger, Lutz-Uwe Glöckner, Dirk Karmann, Karl-Heinz Beuthe; (back to camera) Helmut Binder on the left, unknown on the right. Three months earlier, on 17 March 1965, Kapitänleutnant Karmann had brought the first F-104G for MFG 2 to Eggebek. The F-104G parked in the background (VB+231, serial no. 7188) had been handed over to MFG 2 shortly before. (BArch, Image-F027432-0008/Photographer: Berretty)

The F-104G of MFG 1 VA+106, serial no. 7086, seen here in spring 1967 during a high-altitude transit flight. The Marineflieger Starfighters were all built by MTT at Ingolstadt-Manching or by FIAT at Turin, Italy. (Horst Robitzkat)

Above left: The MFG 1 wing crest had the Wing lettering at the lower edge of the crest until the mid- 1970s. After a redesign, the lettering was to be found at the upper edge of the crest against a yellow background. (Helmut Baumann)

Above right: Korvettenkapitän Kurt Ziebis, OC 2. Staffel/MFG 1, achieved his 2,000th total flying hour on 10 April 1967, primarily flying the Sea Hawk and Starfighter. He was greeted by his squadron pilots after landing at Schleswig-Jagel. To his right is Oberleutnant zur See Jürgen Worms. (Horst Robitzkat)

Above left: Another view of Korvettenkapitän Kurt Ziebis, welcomed after his 2,000th flying hour. Greeting him with a glass of Korn are (from left) Oberleutnant zur See Günter Grosklos, Kapitänleutnant Klaus Stemmler, Oberleutnants zur See Jürgen Worms and Gerd Gossler. (Horst Robitzkat)

Above right: Kapitänleutnant Ingomar 'Ringo' Suhr of MFG 2 achieved his 2,000th total flight hour in spring 1977. (Hans-Joachim du Roi)

A two-ship formation of F-104Gs from MFG 1 in autumn 1967 south of the town of Schleswig in a left turn towards the naval air base. (via Henning Müller-Nagell)

Above left: Flying hours anniversaries are often celebrated in squadron life with a water drenching from the fire brigade. In this case, Kapitänleutnant Ingomar 'Ringo' Suhr was certainly glad to be wearing an immersion suit in the spring of 1977. A few years later, on 14 August 1983, during an open day at Eggebek, he flew the first display of the new official Vikings team as formation leader. His Number Two on that day was Kapitänleutnant Manfred Schulze. (Hans-Joachim du Roi)

Above right: In August 1978, a team from MFG 2 took part in Tactical Fighter Weaponry at Aalborg, Denmark. The aim of this exercise was to train tactical operational procedures in a multi-national framework. Taking part were detachments from operational squadrons of several NATO nations: (on cockpit ladder) Fregattenkapitän Hartmut Fetz (Deputy OC Flying Group); (standing from left) Kapitänleutnants Reinhard Rademacher, Herbert Resch, Rainer Späth, Rainer Mecklenburg, Ingomar 'Ringo' Suhr, Ulrich Korst, Korvettenkapitän Lothar Martin; (lying from left) Kapitänleutnant Dieter-Wolfgang Günter, Korvettenkapitän Hermann Eichin (OC 2. Staffel/MFG 2). The two Starfighters wore the new camouflage paint scheme introduced by the Marineflieger in the summer of 1969, where the upper side was basalt grey with an unchanged lower side of white aluminium. (via Hermann Eichin)

On 23–24 June 1979, the International Air Tattoo, a huge air show, took place at the US air base Greenham Common, UK. One month earlier, the first MFG 2 Starfighter duo had been formed on the occasion of an open day at Eggebek. However, bad weather at Eggebek prevented the team from taking off, so the successful demonstration a few weeks later at Greenham Common was its premiere. Taking off for the flight demonstration on 23 June 1979 are Korvettenkapitän Jürgen Tank (Leader) and his Number Two, Kapitänleutnant Karsten Wichert. After the fatal crash of Kapitänleutnant Manfred Stürmer of MFG 1 during an air display at Royal Naval Air Station (RNAS) Yeovilton, UK, a few weeks later on 3 August 1979, the Starfighter duo project, the forerunner of the later Vikings, was abandoned until 1983. The RF-104G 21+18, serial no. 6673, was the last Starfighter of the Marineflieger to be lost in a crash. On 27 March 1985, its engine failed after launching an AS-30 air-to-ship training missile over the North Sea. Oberleutnant zur See Reinhard Dresbach successfully ejected and was recovered from the North Sea a short time later. (Klaus Kropf)

In August 1979, a large formation of MFG 2, with a total of 12 F-104Gs and one TF-104G, was seen overflying Eggebek. The shadow of the formation created by the sun is clearly visible on the grassy area next to Runway 19, while the aircraft themselves are barely visible in front of the village of Eggebek. Several RAF Buccaneer fighter-bombers are parked next to the aircraft shelters east of the runway. (Hans-Joachim du Roi)

Above left: F-104G 22+14, serial number 7085, of MFG 1 during a stopover at Nörvenich in November 1979. The aircraft was built by MTT at Ingolstadt-Manching and delivered to the wing at Schleswig-Jagel in April 1964. Bearing the registration VA+105 until 1968 it stayed with MFG 1 until October 1980. The special underwing pylon under the left wing was for the air-to-ship missiles AS 30 and Kormoran 1. (Helmut Baumann)

Above right: The Kormoran 1 and AS 30 special underwing pylon. (Helmut Baumann)

Two F-104G of MFG 2 in tactical formation over the Baltic Sea in the summer of 1980. Both aircraft carry BL 755 practice bombs under their wings and AIM-9B Sidewinder air-to-air practice missiles for self-defence on the Aero 3B Catamaran carrier. The forward aircraft with 26+66, serial No. 7412, is one of 50 additional F-104Gs built by MBB at Ingolstadt-Manching in 1971 and 1972. (Hans-Joachim du Roi)

A 16-ship formation – eight F-104Gs each from MFG 1 and MFG 2 – flying over Kiel-Holtenau Naval Air Station (NAS) and the Naval Air Division HQ on 6 July 1980. Shortly afterwards, preparations began at MFG 1 for the Wing's conversion to the new Tornado weapon system. The last Starfighter flight of MFG 1 took place after 18 years, on 29 October 1981, at Schleswig-Jagel. Twenty-five-year-old Oberleutnant zur See Ernst Hansen-Hagge, the Wing´s youngest pilot, was in one of the cockpits on that farewell occasion. (Günter Grondstein)

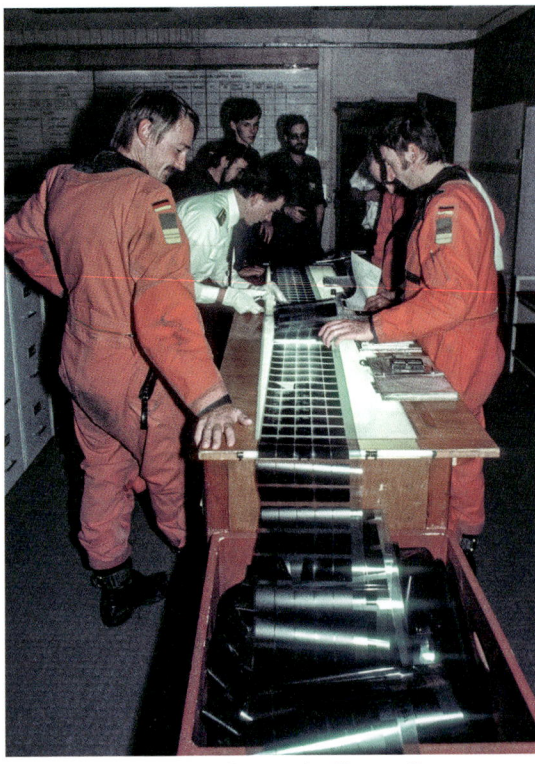

Above left: On 1 June 1979, Korvettenkapitän Günter Grosklos of MFG 1 was the first Marineflieger pilot to achieve 3,000 flying hours on the Starfighter. The first to congratulate him after landing at Schleswig-Jagel was the Commander Naval Air Division, Flottillenadmiral Rudolf Deckert (left). Almost ten years earlier, on 28 August 1970, Kapitänleutnant Günter Grosklos and Oberleutnant zur See Ulrich Otto were the first Marineflieger pilots to bale out from a Starfighter (TF-104G 27+30, serial no. 5732) with Martin-Baker ejection seats, following an engine failure shortly after take-off from Schleswig-Jagel. Both were uninjured. The crash occurred during the first familiarisation flight of Oberleutnant zur See Ulrich Otto at MFG 1 after his successful completion of the Europeanisation course at Jever. (via Günter Grosklos)

Above right: After a reconnaissance flight in September 1982, a film taken during the flight is studied by Photographic Interpreters in the processing room of 1. Staffel/MFG 2. Kapitänleutnant Axel Ostermann (left) and Oberleutnant zur See Siegfried Schmidt wait for the result of the photographic interpretations. (Hans-Joachim du Roi)

In September 1982, 1. Staffel/MFG 2 hosted a squadron exchange with the Dutch F-104G reconnaissance squadron, No. 306 Squadron from Volkel, Netherlands. Pilots from both squadrons gathered for a commemorative photo on a Dutch F-104G, with a recce pod under the fuselage: (on the wing from left) Eerste Luitenant Joost Steenbergen (RNLAF), Tweede Luitenant Ruud Muyrers (RNLAF), Kapitänleutnants Lothar Stryi, Axel Ostermann; (on the fuselage from left) Eerste Luitenant Theo Kustner (RNLAF), Oberleutnants zur See Horst Hartmann, Hagen Haar, Kapitänleutnant Burghard Hachulla, Erste Luitenant Pieter Enzerink (RNLAF), Oberleutnants zur See Hans-Heinrich Surborg, Rolf-Michael Bittner, Korvettenkapitän Jürgen Tank (OC 1. Staffel/MFG 2), Kapitein Dirk Radstake (RNLAF, Commander Dutch Detachment), Kapitänleutnants Hans-Michael Boulanger, Ulli Bauer. (Hans-Joachim du Roi)

Above: 3,000 hours badge.

Right: Fregattenkapitän Hermann Eichin of MFG 2 was one of the very few German pilots who can look back on more than 3,000 hours of Starfighter flying time. He is seen here leaving the cockpit in October 1983 after landing at Eggebek with exactly 3,000 F-104 flight hours. (Hermann Eichin)

The third year of the Vikings display team was 1985. Here, we see them on 28 May 1985 after a final training flight over Eggebek for the display to be held at Prestwick, Scotland, on 6 June 1985. The Leader was Kapitänleutnant Axel Ostermann; the Number Two was Kapitänleutnant Siegfried Schmidt. The Vikings' programme consisted of tactical flight manoeuvres in the horizontal plane. No risky vertical manoeuvres were flown. Both aircraft were from the package of 50 additional Starfighters built by MBB at the beginning of the 1970s. Aircraft 26+60, serial no. 7406, went to the Test Centre at Ingolstadt-Manching after its use by MFG 2. With the registration 98+04, the aircraft undertook the last flight of a German Starfighter on 22 May 1991 (see Volume Two). (Hans-Joachim du Roi)

Above: A reconnaissance flight over the Baltic Sea in May 1985. RF-104G of MFG 2 flies past the chalk cliffs of the Danish island of Møn. (Hans-Joachim du Roi)

Left: On 20 June 1985, the 1. Staffel/MFG 2 could look back on a total of 100,000 flying hours since the squadron was formed with the Sea Hawk in 1958. Standing in front of an RF-104G with the corresponding lettering are (from left) Korvettenkapitän Klaus Pflüger (OC 1. Staffel), Kapitänleutnants Gerd König, Reinhard Breidenbach, Siegfried Schmidt, Oberleutnant zur See Horst Hartmann, Kapitänleutnants Frank Genge; (kneeling from left) Meinhard Aringhoff, Oberleutnants zur See Jan Michels, Reinhard Dresbach. (Hans-Joachim du Roi)

Above: Conversion to the Tornado cast its shadow over MFG 2. Due to necessary construction work, no flight operations were possible at Eggebek from March to September 1986. The aircraft were therefore deployed to Schleswig-Jagel. Seen here, view of the parking area for the MFG 2 Starfighters at Jagel in June 1986. (Klaus Kropf)

Right: The Sea Hawk FGA.6 of the Royal Navy Historic Flight, stationed at RNAS Yeovilton in Somerset, UK, is seen accompanied by two F-104Gs of MFG 2 and two Tornados of MFG 1 during a low-level flight over Schleswig-Holstein. The flight was in celebration of the 50th anniversary of the Schleswig-Jagel AB on 7 July 1985. (Hans-Joachim du Roi)

Above left: In June 1986, the Vikings accepted an invitation to an airshow at Sion, Switzerland. This also resulted in a mixed formation with jets of the Swiss Air Force. Led by a Mirage IIIRS, a Hunter and two F-5E Tiger IIs fly over the mountains near Sion alongside the two Vikings. The Vikings thrilled 120,000 spectators with their display. (Axel Ostermann)

Above right: From 11 to 21 August 1986, a four-ship F-104G formation from MFG 2 flew to the United States on a farewell Starfighter tour. With stopovers on the way there and back in Scotland, Iceland, Greenland and Canada, several USAF and US Navy airfields were visited. A Breguet Atlantic of MFG 5, with technicians and spare parts on board, accompanied the formation. The route across the US also passed New York and, here, three of the Starfighters are seen flying over the Statue of Liberty. (Axel Ostermann)

Above: The Vikings over the Golden Gate Bridge near San Francisco, California. (Axel Ostermann)

Left: For the last air displays of the Vikings in 1986, two aircraft received the blue-white-red colours of Schleswig-Holstein. They were F-104G 26+63, serial no. 7409, and 26+72, serial no. 7418, seen here in close formation over the North Sea coastline in September 1986. (Axel Ostermann)

Above left: During an airshow at Moffett Field, California, the two Vikings pilots, Kapitänleutnant Axel Ostermann and Kapitänleutnant Siegfried Schmidt, thrilled more than 500,000 spectators. Standing in front of one of the large airship hangars, dating from the 1930s, at Moffett Field are the Starfighter pilots on the United States tour: (from left) Kapitänleutnants Harald Bernecker, Siegfried Schmidt, unknown US Navy fan, Fregattenkapitän Jürgen Vollmer, Kapitänleutnant Axel Ostermann. (via Axel Ostermann)

Above right: The vertical stabilisers of the four F-104Gs bear the names of all airfields flown to during the United States farewell tour. (Helmut Baumann)

Above left: The Vikings on Runway 19 at Eggebek. (Axel Ostermann)

Above right: The Dutch air base Eindhoven organised an airshow on 20 September 1986. The Vikings impressed spectators and organisers with their show and received the AGL Trophy for the best jet demonstration from the Aviation Group Leeuwarden. Kapitänleutnants Axel Ostermann (right) and Hagen Haar are seen here after the presentation of the AGL Trophy. On the left is a representative of the Aviation Group Leeuwarden. (Axel Ostermann)

Right: 11 September 1986 marked the official farewell to the Starfighter at Eggebek and the welcoming of the first Tornado to MFG 2. Kapitänleutnants Axel Ostermann (left) and Hagen Haar report prior to the last flight of the Vikings over their home base. (via Axel Ostermann)

Below: The first Tornados fly at MFG 2 – A TF- 104G and a RF-104G in formation with one of the new Tornados. (Axel Ostermann)

At the end of August 1986, the remaining Starfighters of MFG 2 moved to LwVersRgt 1 at Erding. The flying status of the F-104 pilots could thus be maintained until they began their Tornado conversion training. RF-104G 21+26, serial no. 6687, lands at Erding after a low-level flight over southern Germany. The MFG 2 aircraft now also display the emblem of LwVersRgt 1 on the left side of the fin. (Gottfried Schwarz)

An F-104 diamond formation of MFG 2 in autumn 1986 during the time at Erding. (Axel Ostermann)

Above left: The disbandment of the MFG 2 detachment at Erding and the farewell ceremony took place in May 1987. As a farewell gift, OC MFG 2, Kapitän zur See Wolfgang Engelmann (left) handed over part of a Starfighter vertical stabiliser adorned with both coats of arms, to the LwVersRgt 1. Flottillenadmiral Jürgen Dubois (right), Commander of the Naval Air Division, was a guest at the event. The use of the F-104G Starfighter weapon system by the Naval Air Wings had now officially come to an end. (via Klaus Kropf)

Above right: Fregattenkapitän Peter Petersen achieved the most Starfighter flying hours of any Marineflieger pilot. On 2 September 1987 he flew from Erding to Eggebek in a TF-104G of the Luftwaffe and was welcomed there by OC MFG 2, Kapitän zur See Wolfgang Engelmann (right). Fregattenkapitän Petersen achieved 3,757 F-104 hours and thus had flown the second highest number of F-104 hours of all German Starfighter pilots after Oberstleutnant Karl 'Yogi' Söldner (see Volume Two). (Peter Petersen)

Chapter 8
Kommando F-104 – F-104 Holding Unit

Erding Air Base, Bavaria
F-104 in service 1984–88. Unit disbanded.

The Starfighter wings converting to the new Tornado weapon system handed over their Starfighters early to maintenance units to carry out all preparations necessary for a smooth transition to the Tornado. As a result, many pilots who were waiting for their Tornado training at the Tri-National Tornado Training Establishment (TTTE) in the UK were initially given the opportunity to maintain flying currency with other Starfighter units. This also applied to F-104 pilots who were not scheduled for Tornado conversion. To ease the organisation of this operational flying task, a centralised F-104 Holding Unit, the Kommando F-104, was formed in May 1984 at Erding as part of Luftwaffenversorgungsregiment 1 (LwVersRgt 1).

Until September 1988, Kommando F-104 at Erding was, therefore, the temporary flying home for a total of 160 Starfighter pilots of the Luftwaffe and the Marineflieger. About 10,000 flying hours were logged in over four years of its existence.

Right: Kommando F-104 was part of LwVersRgt 1 at Erding. The vertical stabiliser of F-104G 20+37, serial no. 2044, carried a large jubilee emblem to celebrate the 30th anniversary of the regiment in September 1986. The crests of all active and former Starfighter wings of the Luftwaffe and the Marineflieger were applied in a circle around the regimental crest. (Klaus Kropf)

Below: F-104G with the anniversary emblem of LwVersRgt 1 in flight in September 1986. The aircraft was unpainted and polished to a high gloss with white wing upper surfaces as a reminder of the first German Starfighters in the early 1960s. (Gottfried Schwarz)

Left: The first OC of Kommando F-104, Oberstleutnant Gerhard Engelmayer, achieved his 4,000th flying hour in April 1985. At the same time, Major Gottfried 'Blacky' Schwarz (seated) reached 2,000 flying hours. He took over as OC a few months later, in September 1985. (Gottfried Schwarz)

Below: A three-ship formation of F-104Gs in spring 1986 at Erding seen here prior to take-off on Runway 26. They were en route to the nearby Siegenburg air-to-ground firing range. All three aircraft bore the emblem of LwVersRgt 1 on the left side of the stabilisers. The nearest aircraft, with the registration 26+53, serial no. 7313, and previously with JaboG 33, had sported the new all-round camouflage for a good two years. (Gottfried Schwarz)

Hauptmann Wolfgang Leuthner of JaboG 32 shortly before being ready to taxi at Erding for his last F-104 flight before beginning his Tornado conversion training in the UK at the TTTE, RAF Cottesmore in spring 1984. (Wolfgang Leuthner)

Many Starfighters were assigned to LwVersRgt 1 for only a short time after the end of their service with the fighter-bomber wings. These four F-104Gs, previously flown by JaboG 33 and JaboG 34 and seen here in spring 1986, were handed over to the Turkish Air Force between July and October of the same year. During their flying service with Kommando F-104, all four bore the regimental emblem of LwVersRgt 1 on the left side of the fin and the emblem of Technische Gruppe 11 on the right. (Gottfried Schwarz)

The coat of arms of Technische Gruppe 11 was displayed on the right side of the vertical stabilisers of the Starfighters. This Erding-based maintenance unit also provided technical support for the newly formed 'WaKo', the initial Luftwaffe Tornado Weapon Training Unit. (Klaus Kropf)

The Vikings' F-104G 26+63, serial no. 7409, of MFG 2 wore a special blue-white-red livery for its last displays of 1986. Subsequently, it was re-painted in the grey naval camouflage pattern. In April 1988 Major Gottfried 'Blacky' Schwarz ferried the aircraft to Araxos, Greece, for handover to the Greek Air Force. Aircraft 26+63 is seen here during the flight to Greece. (Gottfried Schwarz)

Left: In the final days of Kommando F-104 at Erding, F-104G 22+91, serial no. 7174, wore a dazzling special silver livery. (Gottfried Schwarz)

Below: On 19 September 1988, Major Gottfried 'Blacky' Schwarz flew his last F-104 flight in 22+91. At the same time, this was the official last flight of Kommando F-104, which was disbanded at the end of the month: (from left) Majors Eckehard Wutke, Gottfried 'Blacky' Schwarz, Hauptmann Werner Nemetschek, Major Peter Reimers. (Gottfried Schwarz)

Abbreviations

AB	Air Base
AG	Aufklärungsgeschwader, 'I' for Immelmann – Reconnaissance Wing
AGL	Aviation Group Leeuwarden
AFB	Air Force Base
AIM	Air Intercept Missile
ARGE	Arbeitsgemeinschaft – Working Group
ATAF	Allied Tactical Air Force
ATC	Air Traffic Control
CAT	Category
CRC	Control and Reporting Centre
FAC	Forward Air Controller
FIAT	Fabbricia Italiana Automobili di Torino
FN	Fabrique Nationale
HQ	Headquarters
IWM	Imperial War Museum
JaboG	Jagdbombergeschwader, 'B' for Boelcke – Fighter-Bomber Wing,
JG	Jagdgeschwader, 'R' for Richthofen 'M' for Mölders – Fighter Wing
KPDSU	Kommunistische Partei der Sowjetunion – Communist Party of the Soviet Union
LwVersRgt	Luftwaffenversorgungsregiment – Luftwaffe Supply Regiment
MAP	Military Assistance Programme
MB	Martin-Baker
MBB	Messerschmitt-Bölkow-Blohm
MFG	Marinefliegergeschwader – Naval Air Wing
Mod	modified
MTT	Messerschmitt
NAS	Naval Air Station
NASARR	North American Search and Ranging Radar
OC	Officer Commanding
QRA	Quick Reaction Alert
RNAS	Royal Naval Air Station
SABCA	Sociétés Anonyme Belge de Constructions Aéronautiques
SACA	Società per Azioni Costruzioni Aeronautiche
SATS	Short Airfield for Tactical Support
RNLAF	Royal Netherlands Air Force
SACA	Società per Azioni Costruzioni Aeronautiche
TCTP	Tactical Combat Training Programme
TTTE	Tri-National Tornado Training Establishment
USAF	United States Air Force
VFW	Vereinigte Flugtechnische Werke
WaKo	Waffenausbildungskomponente – Tornado Weapon Training Unit
WaSLw	Waffenschule der Luftwaffe – Luftwaffe Weapon School

German Ranks – Royal Air Force/Royal Navy

Feldwebel	Sergeant
Flottillenadmiral	Commodore
Fregattenkapitän	Commander
Generalleutnant	Air Marshal
Generalmajor	Air Vice Marshal
Gefreiter	Leading Aircraftman – Air Specialist (Class 2)
Hauptfeldwebel	no equivalent
Hauptmann	Flight Lieutenant
Kapitänleutnant	Lieutenant
Kapitän zur See	Captain
Korvettenkapitän	Lieutenant Commander
Leutnant	Pilot Officer
Leutnant zur See	Commissioned Warrant Officer
Major	Squadron Leader
Oberfeldwebel	Flight Sergeant
Obergefreiter	Senior Aircraftman – Air Specialist (Class 1)
Oberleutnant	Flying Officer
Oberleutnant zur See	Sub-Lieutenant
Oberst	Group Captain
Oberstabsarzt	Squadron Leader Medical Doctor
Oberstleutnant	Wing Commander
Stabsarzt	Flight Lieutenant Medical Doctor
Stabsfeldwebel	Warrant Officer
Stabsunteroffizier	Corporal

Bibliography

Bashow, David, *Starfighter: A Loving Retrospective of the CF-104 Era in Canadian Fighter Aviation 1961-1986*, Fortress Publications, Inc., Stoney Creek, Ontario (1990)

Flight Manual F/RF/TF-104G (GAF T.O. 1F-104G-1)

Fischbach, Georg, *Starfighter/916 German F-104/Construction and Life Stories*, Holzkirchen (1996)

Kropf, Klaus, *Deutsche Starfighter*, JOMO Medien-Service: Cologne (1994)

Niccoli, Riccardo, *Coccarde Tricolori Speciale 2: F-104S*, RN Publishing, Novara (2007)

Niccoli, Riccardo, *Coccarde Tricolori Speciale 4: F/TF/RF-104G*, RN Publishing, Novara (2010)

Publications to date

Kropf, Klaus, *Deutsche G.91*, 201 Bücher-Service: Fürstenfeldbruck (2017)

Kropf, Klaus, *Deutsche Starfighter*, JOMO Medien-Service: Cologne (1994)

Kropf, Klaus, *German Starfighters*, Midland Counties Publications, Hinckley (2002)

Kropf, Klaus, *Jet-Geschwader im Aufbruch*, VDM: Zweibrücken (2005)

Other books you might like:

Historic Military Aircraft
Series, Vol. 9

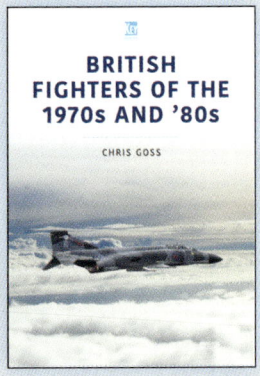

Historic Military Aircraft
Series, Vol. 2

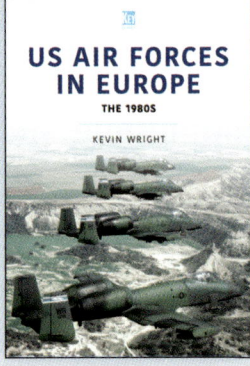

Air Forces
Series, Vol. 4

Historic Military Aircraft
Series, Vol. 15

Historic Military Aircraft
Series, Vol. 17

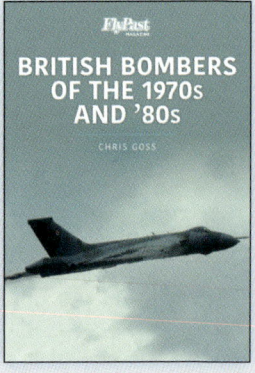

Historic Military Aircraft
Series, Vol. 4

For our full range of titles please visit:
shop.keypublishing.com/books

VIP Book Club

Sign up today and receive
TWO FREE E-BOOKS

Be the first to find out about our forthcoming
book releases and receive exclusive offers.

Register now at **keypublishing.com/vip-book-club**

Our VIP Book Club is a 100% spam-free zone, and we will never share your email with anyone else.
You can read our full privacy policy at: privacy.keypublishing.com